Fungicolous, Lichenicolous, and Myxomyceticolous Species of *Hypocreopsis*, *Nectriopsis*, *Nectria*, *Peristomialis*, and *Trichonectria*

Gary J. Samuels

NYBG

The New York Botanical Garden
Bronx, New York 10458
Issued: 5 October 1988

MEMOIRS OF THE NEW YORK BOTANICAL GARDEN VOLUME 48

Published by
The New York Botanical Garden
Bronx, New York 10458

International Standard Serial Number 0071-5794

Library of Congress Cataloging-in-Publication Data

Samuels, Gary J.
Fungicolous, lichenicolous, and myxomyceticolous species of Hypocreopsis, Nectriopsis, Nectria, Peristomialis, and Trichonectria.

(Memoirs of the New York Botanical Garden ; v. 48)
"Issued September 1988."
Bibliography: p.
Includes index.
1. Hypocreaceae—Classification. 2. Fungicolous fungi—Classification. 3. Lichenicolous fungi—Classification. 4. Myxomyceticolous fungi—Classification. 5. Hypocreaceae—Identification. 6. Fungicolous fungi—Identification. 7. Lichenicolous fungi—Identification. 8. Myxomyceticolous fungi—Identification. 9. Fungi—Classification. 10. Fungi—Identification. I. Title. II. Series.
QK1.N525 vol. 48 581 s 88-28896
[QK623.H86] [589.2′3]
ISBN 0-89327-330-9

International Standard Book Number 0-89327-330-9

OCTOBER 1988 MEMOIRS OF THE NEW YORK BOTANICAL GARDEN **48:** 1–78

Fungicolous, Lichenicolous, and Myxomyceticolous Species of *Hypocreopsis, Nectriopsis, Nectria, Peristomialis,* and *Trichonectria*

Gary J. Samuels[1]

Contents

Summary

Samuels, Gary J. (The New York Botanical Garden, Bronx, New York 10458). Fungicolous, lichenicolous, and myxomyceticolous species of *Hypocreopsis, Nectriopsis, Nectria, Peristomialis,* and *Trichonectria.* Mem. New York Bot. Gard. **48:** 1–78. 1988. This is an account of hypocrealean fungi that have white to pale yellow or orange perithecia and that, for the most part, occur on ascomycetes. Included here also are related species that occur on lichens or myxomycetes and, while not obviously fungicolous, occur on substrates that are colonized by other fungi and thus may be fungicolous. Sixty species in the genera *Hypocreopsis, Nectria, Nectriopsis, Peristomialis,* and *Trichonectria* are discussed. The genera *Nectriopsis* and *Peristomialis* are removed from synonymy with *Nectria. Bionectria* is relegated to synonymy with *Nectria* and the type species is included in *Nectria.* One new species of *Hypocreopsis* is described. Four species, one new, are included in *Nectria. Nectriopsis* is

[1] The New York Botanical Garden, Bronx, New York 10458.

emended to include species that have white to pale yellow perithecia, a perithecial wall less than 20 μm wide, once-septate ascospores and, generally, a fungicolous habit. Forty-three species of *Nectriopsis* are described or are redescribed. Twenty-seven new combinations and 18 new species are proposed. A key to the genera of the Hypocreales known to include fungicolous, lichenicolous, or myxomyceticolous species is included. Keys to fungicolous, lichenicolous, and myxomyceticolous species in the genera *Hypocreopsis, Nectria, Nectriopsis, Peristomialis,* and *Trichonectria* are presented. Descriptions are provided for all species and illustrations are provided for those species that have not previously been illustrated.

Introduction

In the present work 60 species of *Hypocreopsis* Karsten, *Nectria* (Fries) Fries, *Nectriopsis* Maire, *Peristomialis* (Phillips) Boudier, and *Trichonectria* Kirschstein are redescribed or are described as new. The majority of these species occur on ascomata of other ascomycetes, frequently on perithecia of members of the Hypocreales. A few are not unequivocally fungicolous but are found on substrates that are colonized by other fungi and, so, may be hyperparasitic on mycelium.

These fungi are Ascomycetes, members of the Hypocreales. Their perithecia are minute, usually less than 200 μm in diameter, and inconspicuous. They are white to yellow and do not become red in potassium hydroxide. These species are not closely related to *Nectria episphaeria* (Tode: Fries) Fries and its allies, fungicolous species that have red perithecia and that are the most commonly collected of the fungicolous nectrias (see Booth, 1959; Dennis, 1978).

This study is an account of specimens that I have encountered while studying the Hypocreales primarily found in neotropical areas. It was surprising to find so many species on such a seemingly specialized substrate as other pyrenomycetes and no doubt many other hyperparasitic species are yet to be found. The work is thus preliminary. Most of the species included in this study are known only from single collections. The limits of individual species may thus not yet be clear. The generally fungicolous habit and tiny, fleshy, yellow to white perithecia suggest a taxonomic homogeneity that is more apparent than real. The variation in perithecial anatomy, and to a lesser extent anamorphs—features given high taxonomic significance (Booth, 1959; Rossman, 1983; Samuels & Rossman, 1979; Samuels & Seifert, 1987)—indicate taxonomic diversity. Furthermore, many non-hypocrealean ascomycetes that occur on other ascomycetes have fleshy, pale yellow ascomata. These include, among others, the inoperculate discomycete *Calloriopsis gelatinosa* (Sacc.) Sydow (Pfister, 1976) and several genera of ascomycetes the species of which have bitunicate asci (e.g., *Letendraea* Sacc., *Melioliphila* Speg., *Tubeufia* Penzig & Sacc.; see Barr, 1980; Pirozynski, 1976; Rossman, 1987). Thus yellow coloration is a generalized adaptation to a fungicolous habit.

Twenty-four genera currently attributed to the Hypocreales are known to have one or more fungicolous, lichenicolous, or myxomyceticolous species. Most of these genera are poorly known, and at least some may not actually belong to the order. A key, adapted from Rogerson (1970), to the genera of the Hypocreales that have fungicolous, lichenicolous, or myxomyceticolous species follows.

Materials and Methods

Ascospores were isolated on cornmeal dextrose agar (CMD, Difco) with the aid of a micromanipulator. Colonies were studied on CMD, potato dextrose agar (PDA, Difco) and oatmeal agar (OA, Difco). Herbarium specimens were rehydrated in 3% KOH and subsequently studied in water, 3% KOH, 100% lactic acid, or cotton blue in lactic acid. Citation of localities of collections made in New Zealand follows Crosby et al. (1976).

Key to Genera of the Hypocreales Known to Have Fungicolous, Lichenicolous, and Myxomyceticolous Species

(The genera indicated by Roman numerals are included in this study.)

1. Perithecia immersed more or less completely in a compact stroma; stromata superficial or erumpent through substrate; subiculum lacking.
   2. Ascospores separating into part spores; asci cylindrical, 8-spored in early stages, at maturity appearing 16-spored; ascospores hyaline or green; stromata various, effuse and thin, forming a crust, to globose, pulvinate to cupulate. *Hypocrea* Fries.
   2. Ascospores not separating into part spores; asci clavate, 8-spored; ascospores hyaline; stromata pulvinate to tuberculate. III. *Hypocreopsis.*
1. Perithecia not immersed in a compact stroma, either immersed in the substrate and separate, or cespitose and superficial on the substrate; with or without a subiculum.
   3. Perithecia immersed in the substrate, finally more or less erumpent, at least the ostiolar region; with or without a basal stroma, with or without a subiculum.
      4. Ascospores 1-celled, ornamented with a network of bands forming a narrow border; perithecia immersed in the hymenium of host (*Tuber*), appearing to be nonostiolate. *Battarina* (Sacc.) Clements & Shear.
      4. Ascospores more than 1-celled, not ornamented with bands, not on *Tuber*; perithecia ostiolate.
         5. Ascospores 2-celled.
            6. Ascospores colorless, >5 μm long; asci 8-spored; perithecia immersed in lichen thallus. *Nectriella* Fuckel.
            6. Ascospores dark, 5 μm or less long; asci becoming many-spored; not lichenicolous. *Erispora* Patouillard.
         5. Ascospores more than 2-celled.
            7. Ascospores with transverse septa only.
               8. Ascospores brown, 4-celled; perithecia globose, occurring in empty larger perithecia of host (*Balansia*). *Weesia* Höhnel.
               8. Ascospores hyaline, 2–4-celled; perithecia globose to egg-shaped; not occurring on *Balansia.*
                  9. Perithecia long egg-shaped with projecting ostiolar region, hyaline to yellowish; asci cylindrical, 8-spored; ascospores spindle-shaped, with 3 distinct septa; on *Valsa.* *Debaryella* Höhnel.
                  9. Perithecia globose with small, flattened ostiolar region, hyaline to yellowish; asci cylindrical or clavate, 8-spored; ascospores spindle-shaped to vermiform with 1–3 indistinct septa; on *Leptosphaeria.* *Cryptoleptosphaeria* Höhnel.
            7. Ascospores at first 2-celled then becoming falsely multiseptate to muriform with both transverse and longitudinal septa, cylindrical, ellipsoidal to oblong; asci 2(3–5)-spored; perithecia semi-immersed; hyaline to fuscous, warted; lichenicolous. *Xenonectriella* Weese.
   3. Perithecia superficial from the start, with or without a subiculum or basal stroma.
      10. Perithecia without a subiculum or basal stroma.
          11. Ascospores 1-celled.
              12. Perithecia pale yellow with long cylindrical ostiolar region, fimbriate at the mouth; asci globose to broad-clavate, 8-spored, deliquescing. *Sphaeronaemella* Karsten.
              12. Perithecia concave to globose, papillate, not rostrate; asci clavate to cylindrical, not deliquescing.
                  13. Asci with more than 8 spores; ascospores ovate, hyaline; lichenicolous. *Ahlesia* Fuckel.
                  13. Asci 8-spored or less; ascospores short acicular, ellipsoidal or subellipsoidal; not on lichens.
                      14. Ascospores short acicular; perithecia fleshy gelatinous, gregarious to cespitose; asci clavate cylindric; on *Hypocrella.* *Epicrea* H. & P. Sydow.
                      14. Ascospores ellipsoidal or subellipsoidal; perithecia soft fleshy or membranous; asci cylindrical; not on *Hypocrella.* *Pseudonectria* Sever.
          11. Ascospores more than 1-celled.
              15. Ascospores 2-celled.
                  16. Perithecia rostrate; ascospores ciliate or not at one or both ends, sometimes surrounded by a layer of mucus; on myxomycetes. *Rhynchonectria* Höhnel.
                  16. Perithecia papillate or nonpapillate, not rostrate; ascospores nonciliate.
                      17. Perithecia blue or violet; asci 8-spored; ascospores remaining whole. *Lisea* Sacc.
                      17. Perithecia not blue or violet, typically white, yellow, orange, red, or occa-

sionally brown; asci remaining 8-spored or becoming 16-spored, ascospores separating at the septum into part spores.

18. Asci becoming 16-spored.
    19. Perithecia orange to olivaceous or black with rigid hairs. *Neorehmia* Höhnel.
    19. Perithecia typically red, hairs lacking. *Neoskofitsia* Schulzer.
18. Asci remaining 8-spored.
    20. Perithecia minute, yellow; perithecial wall <15 μm wide, of a single region of cells; occurring on ascomycetes, often on *Nectria*. V. *Nectriopsis*.
    20. Perithecia larger, yellow or orange to red; perithecial wall >20 μm wide, of two or more regions of cells; on stromatic ascomycetes but not on Hypocreales. IV. *Nectria*.

15. Ascospores more than 2-celled.
    21. Ascospores ciliate at each end, hyaline, mostly 4-celled but sometimes 2–7-celled; cylindrical; perithecia lacking distinctive hairs. *Paranectria* Sacc.
    21. Ascospores not ciliate, $\geq$3-celled; perithecia with or without distinctive hairs.
        22. Perithecial apex not discoidal, not formed of intertwined hyphae; spine-like, thick-walled setae scattered over the surface of the perithecial wall. I. *Trichonectria*.
        22. Perithecial apex discoidal, formed of intertwined hyphae; hairs lacking, or tooth-like, hyphal hairs formed of a fascicle of hyphae arising from the perithecial apex. II. *Peristomialis*.

10. Perithecia in or on a subiculum, or with a basal stroma.
    23. Perithecia nonstromatic, in or on a subiculum.
        24. Perithecia rostrate with awl-shaped beak; asci clavate, deliquescing; ascospores smooth-walled, fusiform. *Pyxidiophora* Brefeld & Tavel.
        24. Perithecia globose to papillate, not rostrate; asci cylindrical, persistent; ascospores smooth, spinulose or warted, globose, ellipsoidal, naviculate or fusiform.
            25. Asci appearing to be 16-spored; ascospores disarticulating at the septum. *Hypocrea* Fries.
            25. Asci 8-spored, ascospores remaining 2-celled.
                26. Ascospores ellipsoidal to fusiform, not apiculate, unicellular, hyaline, smooth or warted; perithecia white, yellow or violaceous (one species); on Fungi Imperfecti, ascomycetes or myxomycetes. V. *Nectriopsis*.
                26. Ascospores slipper-shaped, naviculate, or fusiform, apiculate or not, with 0–3 septa, typically warted, sometimes smooth; perithecia white, yellow, orange, or red; on basidiomycetes, rarely on ascomycetes.
                    27. Ascospores 1-septate, ellipsoidal to naviculate, nonapiculate, smooth to spinulose but not warted; perithecia usually superficial on a thin subiculum, scattered and remaining discrete; anamorph *Gliocladium*; occurring on Aphyllophorales. *Sphaerostilbella* (Henn.) Sacc.
                    27. Ascospores 1–2-septate, sometimes apiosporous, slipper-shaped, ellipsoidal, or fusiform, ends acute and often apiculate; perithecia usually immersed in dense mycelium, often densely gregarious to cespitose; anamorphs hyphomycetous and mononematous but not *Gliocladium*, occurring on Aphyllophorales, Agaricales, Boletales, Leotiales (Leotiaceae, Geoglossaceae), Pezizales. *Hypomyces* (Fr.) L.-R. & C. Tulasne.
    23. Perithecia on a basal stroma, the lower portion often immersed in the stroma, subiculum lacking.
        28. Perithecia red or becoming red in 3% KOH. IV. *Nectria*.
        28. Perithecia yellow, not becoming red in 3% KOH. III. *Hypocreopsis*.

Keys to *Nectriopsis* Species, and to Fungicolous, Lichenicolous, and Myxomyceticolous Species of *Hypocreopsis, Nectria, Peristomialis,* and *Trichonectria* that Have White to Yellow Perithecia

Dichotomous Key

1. On myxomycetes.
    2. Perithecia violet to purple. 41. *Nectriopsis violacea*.

2. Perithecia colorless, white, yellow or yellow-brown.
3. Ascospores averaging 9–13 × 4–5 μm. 43. *Nectriopsis hirsuta.*
3. Ascospores averaging <10 μm long.
4. Ascospores (7–)8–9(–10) × 2–3.5 μm; anamorph *Gliocladium.* 42. *Nectriopsis sporangiicola.*
4. Ascospores shorter; anamorph *Acremonium.*
5. Aculeolate, spinulose processes arising from the surface of the perithecial wall; ascospores (4–)4.7–5.5(–7) × 2–3 μm, smooth. 39. *Nectriopsis exigua.*
5. Perithecial wall covered with hyphae but without distinct aculeolate processes; ascospores smooth or spinulose, larger.
6. Ascospores distinctly spinulose, lacking hyaline appendages, (4.5–)5.3–6.7(–8) × (2.5–)2.8–3.4(–4) μm. 38. *Nectriopsis oropensoides.*
6. Ascospores (5–)6–7.5(–8.7) × 2–3 μm, smooth, with a colorless, cap-like appendage at each end while in the ascus. 40. *Nectriopsis candicans.*
1. On substrates other than myxomycetes.
7. On discomycetes or on meliolaceous leaf inhabiting fungi.
8. On meliolaceous leaf inhabiting fungi.
9. Ascospores (14–)16.8–22.2(–27) × 3–4 μm, striate. 31. *Nectriopsis oidioides.*
9. Ascospores smaller, smooth, spinulose or striate.
10. Ascospores averaging >12 μm long.
11. Ascospores striate, (11–)12.4–16.3 × 3–4 μm; perithecial hairs lacking. 32. *Nectriopsis leucorrhodina.*
11. Ascospores smooth; perithecia associated with an *Acremonium* anamorph with thick-walled conidiophores; perithecial hairs present or lacking.
12. Ascospores (8.5–)11.4–15.5(–18) × 2.3–4 μm; ascal apex simple; with hyphal hairs arising from around the perithecial apex. 29. *Nectriopsis cephalosporii.*
12. Ascospores (13.5–)14–16(–17) × 2–3.5(–4) μm; ascal apex with a small ring; perithecial hairs lacking. 30. *Nectriopsis sensitiva.*
10. Ascospores averaging ≤10 μm long.
13. Ascospores 8–10.5 × 2.5–3 μm, smooth to finely spinulose; perithecia lacking modified cells around the apex. 33. *Nectriopsis guarapiensis.*
13. Ascospores 9–11 × 3–4 μm, smooth; with modified, hair-like cells arising from around the perithecial apex. 34. *Nectriopsis pipericola.*
8. On discomycetes.
14. Ascospores (23.5–)24.3–27.7(–29) × (10–)10.6–12.7(–13) μm; on *Sepultaria arenicola.* 20. *Nectriopsis sepultariae.*
14. Ascospores shorter.
15. Ascospores (11–)11.6–14.2(–16) × (6–)6.7–7.7(–8) μm, pale brown, verrucose; on *Bioscypha* sp. on fern (*Lophosoria quadripinnata*). 22. *Nectriopsis discicola.*
15. Ascospores averaging <3 μm wide.
16. Ascospores (10–)11.6–13.5(–14.5) × 2–3 μm, smooth; on an inoperculate discomycete. 3. *Trichonectria albidopilosa.*
16. Ascospores 11–20 × 2–3 μm, or smooth; on *Lachnum* sp. 21. *Nectriopsis discophila.*
7. On lignicolous, corticolous or bambusicolous pyrenomycetes or loculoascomycetes or their anamorphs or mycelium, or on lichens; rarely foliicolous and then not on meliolaceous fungi, sometimes not conspicuously fungicolous.
17. Ascospores averaging ≤10 μm long.
18. Ascospores averaging ≤3 μm wide.
19. Ascospores conspicuously striate, brown. 13. *Nectria brunneostriata.*
19. Ascospores smooth or spinulose, colorless or brown.
20. Perithecial wall surface of intertwined hyphae; perithecial apex with a disc of agglutinated hyphae that may grow outwardly to form a crown of agglutinated, fasciculate hairs.
21. Ascospores brown. 13. *Nectria brunneostriata.*
21. Ascospores colorless.
22. Fasciculate triangular hairs arising from the perithecial apex; on a lichen. 35. *Nectriopsis porinicola.*
22. Perithecia with a disc of agglutinated hyphae, hyphal hairs lacking; on bamboo. 7. *Peristomialis leucocarpa.*
20. Perithecial wall surface of pseudoparenchyma or *textura epidermoidea*; if wall covered with hyphae, then hyphae not agglutinated to form a disc or hairs.
23. Perithecia partially to wholly immersed in mycelium.
24. On immersed ascomycete (Amphisphaeriaceae) on palm; Indonesia. 47. *Nectriopsis epimyces.*

24. On substrates other than palms; Brazil, Japan, New Zealand.
  25. On Hypocreales, mycelium not restricted to fungal host; Brazil. 48. *Nectriopsis cupulata.*
  25. Not conspicuously fungicolous; Venezuela, Japan, New Zealand.
    26. Ascospores (7–)8.5–10.5(–12) × 2.7–3.5 μm, mycelium radiating; Japan, New Zealand. 46. *Nectriopsis septofusidiae.*
    26. Ascospores (5.5–)6–7(–7.5) × 1.6–1.8(–2) μm, mycelium forming dense, rectangular spots; Venezuela. 45. *Nectriopsis nanocarpa.*

23. Perithecia at most seated on a thin mycelium or mycelium not visible.
  27. Ascospores averaging >8 μm long.
    28. Surface of perithecial wall obscured by white to golden hyphae; ascospores oblong, (6–)7.5–10.3(–14) × 1.5–2.3(–3) μm. 37. *Nectriopsis squamulosa.*
    28. Cells at surface of perithecial wall *textura epidermoidea* to pseudoparenchymatous; perithecia associated with white mycelium or mycelium not visible; ascospores ellipsoidal, 5–12.5 × (1.5–)2–3 μm.
      29. Perithecia with circular cells at the surface, cells sometimes joined in chains to form hairs; ascospores (5–)6.7–10.6(–12.5) × (1.5–)2.2–3 μm. 18. *Nectriopsis mindoensis.*
      29. Perithecial wall surface *textura epidermoidea,* hairs not formed; ascospores (7.2–)7.9–9.2(–10) × 3 μm. 57. *Nectriopsis lilliputia.*
  27. Ascospores averaging ≤8 μm long.
    30. Perithecial papilla formed of hyphal elements; hyphal hairs sometimes growing outwardly for a short distance (≤20 μm) to form short hairs; cells at surface of perithecial wall forming *textura epidermoidea,* or angular in outline.
      31. On stromata of members of the Xylariaceae and Hypocreaceae; perithecia averaging ≥140 μm diam.; ascospores (5–)5.7–7.3(–9) × 2–2.5(–3) μm. 19. *Nectriopsis perpusilla.*
      31. Perithecia not conspicuously fungicolous, scattered on dead bamboo and possibly on black mycelium; perithecia averaging <100 μm diam.; ascospores (6.3–)7–8.7(–9) × 2–3.2 μm. 44. *Nectriopsis microthecia.*
    30. Perithecial papilla not formed of distinct hyphal elements, hyphal hairs not growing out from perithecial apex but sometimes with short chains of cells forming cellular hairs; cells at surface of wall forming *textura epidermoidea* or elliptic to circular in outline.
      32. Cells at surface of perithecial wall elliptic to circular in outline, sometimes held in chains to form hairs; perithecial wall 15–20 μm wide; ascospores (5–)6.7–10.6(–12.5) × (1.5–)2.2–3 μm. 18. *Nectriopsis mindoensis.*
      32. Cells at surface of perithecial wall *textura epidermoidea*; perithecial wall ≤15 μm wide, of a single region of cells.
        33. Perithecial papilla of hyphal elements; ascospores (5–)5.7–7.5(–9) × 2–2.5(–3) μm. 19. *Nectriopsis perpusilla.*
        33. Perithecial papilla of small, indistinct cells; ascospores (4.5–)5–6.3(–7.2) × (1.5–)1.8–2.7(–3) μm. 55. *Nectriopsis epimycota.*

18. Ascospores averaging >3 μm wide.
  34. Perithecia forming on stromata of members of the Clavicipitales on grass culms. 59. *Nectriopsis epichloe.*
  34. Perithecia not forming on stromata of members of the Clavicipitales on grass culms.
    35. Ascospores (7–)9–10.5(–11.7) × (4–)4.3–5.5(–6) μm, finely spinulose. 49. *Nectriopsis bactridioides.*
    35. Ascospores averaging ≤5 μm wide, smooth, striate or spinulose.
      36. Ascospores striate.
        37. Ascospores 5.5–6.5(–7.2) × (3–)3.5–3.8(–4) μm; perithecia immersed in ferrugineous mycelium. 54. *Nectriopsis epinectria.*
        37. Ascospores >7 μm long, mycelium white or not visible.
          38. Perithecia seated in or on white mycelium, ascospores (6.3–)8.4–10.5(–11) × 3.5–4(–4.2) μm, striate. 52. *Nectriopsis albofulta.*
          38. Mycelium not visible or scant, ascospores (8–)8.5–11(–12.5) × (3–)3.8–4.7(–5) μm, striate. 26. *Nectriopsis sororicola.*
      36. Ascospores smooth or spinulose.
        39. Ascospores averaging ≤4 μm wide.

40. Ascospores (4.5–)5.3–6.7(–8) × (2.5–)2.8–3.4(–4) μm, spinulose. 38. *Nectriopsis oropensoides.*
40. Ascospores averaging >9 μm long, smooth.
41. Perithecia partially to completely immersed in white mycelium, associated with penicillately branched conidiophores; ascospores (7–)9.5–13(–14) × (3–)3.5–4(–5) μm; on *Nectria.* 50. *Nectriopsis byssotecta.*
41. Mycelium not visible; anamorph *Acremonium*; ascospores (8–)9–11(–12.5) × 3–4 μm; on *Nectria.* 25. *Nectriopsis lasiodermopsis.*
39. Ascospores averaging >4 μm wide.
42. Ascospores (10–)11–12.5(–13) × 4.5–5.5(–6) μm; perithecial wall 50–80 μm wide; perithecia easily crushed in water on a microscope slide by the weight of a coverslip. 56. *Nectriopsis flavella.*
42. Ascospores averaging ≤10 μm long; perithecial wall ≤30 μm wide; perithecia not crushed by the weight of a coverslip.
43. Ascospores (8–)9–11 × 4–5 μm, smooth; on immersed ascomycete. 36. *Nectriopsis hypocrellicola.*
43. Ascospores (7–)7.8–9.6(–10) × 4–4.5(–5) μm, spinulose; on *Nectria.* 27. *Nectriopsis sibicola.*

17. Ascospores averaging >10 μm long.
44. Ascospores averaging ≤3 μm wide.
45. Perithecia wholly to partially immersed in white mycelium; on immersed ascomycete on palm; ascospores (10–)10.5–13(–16.2) × (2–)2.3–3(–3.5) μm, smooth. 47. *Nectriopsis epimyces.*
45. Perithecia at most seated on thin, white to yellow mycelium; on immersed or superficial ascomycetes.
46. Ascospores averaging <12 μm long.
47. Perithecial apex conspicuously formed of intertwined hyphae; ascospores 6–8 × 3–4 μm, finely spinulose. 8. *Peristomialis dentifera.*
47. Perithecial wall not formed of intertwined hyphae; ascospores averaging <3 μm wide, smooth or finely spinulose.
48. Ascospores oblong to narrowly ellipsoidal, (6–)7.5–10.3(–14) × 1.5–2.3(–3) μm; ascal apex with a refractive ring; perithecia seated on white to golden mycelium. 37. *Nectriopsis squamulosa.*
48. Ascospores ellipsoidal, (5–)6.7–10.6(–12.5) × (1.5–)2.2–3 μm; ascal apex simple; perithecia associated with white mycelium or mycelium not visible. 18. *Nectriopsis mindoensis.*
46. Ascospores averaging >12 μm long.
49. Ascospores 14.5–20 × (2.5–)3–5(–5.4) μm, finely spinulose; perithecial apex of intertwined hyphae; on immersed ascomycete on leaves of *Podocarpus* sp. 5. *Peristomialis parilis.*
49. Ascospores shorter, smooth; perithecial apex not of intertwined hyphae; not on leaves of *Podocarpus* sp.
50. Perithecia distinctly discoidal and thick-walled (>15 μm), cupulate when collapsed; ascospores (10–)12.2–14.8(–15) × (2.5–)2.8–3.2(–3.5) μm, smooth; on Xylariaceae. 17. *Nectria discoidea.*
50. Perithecia globose to pyriform and thin-walled (<15 μm wide); ascospores 13–15 × 2–3 μm, smooth; on *Xylaria* and other fungi. 23. *Nectriopsis ostiolorum.*

44. Ascospores averaging >3 μm wide.
51. Perithecial apex and wall conspicuously formed of intertwined hyphae.
52. Ascospores didymosporous or phragmosporous.
53. Perithecial apex with long, radiating, fasciculate hyphal hairs; ascospores fusiform, (18–)25–32(–37) × (3–)4–6(–7) μm, 1–5-septate, striate. 10. *Peristomialis berkeleyi.*
53. Perithecia lacking hyphal hairs.
54. Ascospores (19.5–)21.5–24.5(–25.5) × 4–5 μm, 1-septate, striate. 6. *Peristomialis paraparilis.*
54. Ascospores 27–42 × 3–4.5 μm, 5–13-septate, smooth. 9. *Peristomialis corynospora.*
52. Ascospores dictyosporous. 11. *Peristomialis dictyospora.*
51. Perithecial apex and wall not conspicuously hyphal.
55. Perithecia completely immersed in a rust-colored, cushion-shaped stroma; asco-

spores 16–19 × 7.5–8.5 μm.
Hypocreopsis episphaeria (Pat.) Müller (Müller & Arx, 1962).

55. Perithecia not immersed in a rust-colored, cushion-shaped stroma; ascospores shorter or longer and averaging <7 μm wide.
 56. Ascospores averaging >20 × >4.5 μm.
  57. Ascospores (16–)19.7–24.5(–27) × (5–)5.5–6.5(–7) μm, warted; on foliicolous ascomycetes. 14. *Nectria tonduzi.*
  57. Ascospores (26–)27.8–30.6 × 4.5–5.7(–6) μm, smooth, on *Xylaria.* 12. *Hypocreopsis xylariicola.*
 56. Ascospores averaging <20 × <4.5 μm.
  58. Perithecia partially to completely immersed in white mycelium.
   59. On members of the Clavicipitales on grass culms.
    60. Ascospores (8.5–)9.3–11.5(–13) × (3–)3.3–4(–4.5) μm. 59. *Nectriopsis epichloe.*
    60. Ascospores 15.3–19.2(–23) × 3–4 μm. 60. *Nectriopsis macroepichloe.*
   59. Not on members of the Clavicipitales, not on grasses.
    61. Perithecia with thick-walled setae, or with hyphal hairs; ascospores disarticulating at the septum or not; foliicolous or lignicolous.
     62. Perithecia with stiff, thick-walled setae; ascospores (12.5–)13.6–16.5(–17) × (2.5–)2.8–3.6(–4) μm, disarticulating; foliicolous. 4. *Trichonectria erythroxylifoliae.*
     62. Perithecia with triangular hairs formed of fascicles of hyphae; ascospores (10–)12–17(–22) × 4–5 μm; fungicolous, herbicolous, lignicolous. 16. *Nectria suffulta.*
    61. Perithecia lacking setae or hyphal hairs; ascospores remaining entire; foliicolous or not.
     63. Foliicolous on meliolaceous fungi; ascospores 16.8–22.2 (–27) × 3–4 μm. 31. *Nectriopsis oidioides.*
     63. Not foliicolous, not on meliolaceous fungi; ascospores shorter.
      64. Ascospores striate.
       65. Mycelium not visible; perithecia globose, becoming cupulate, wall 30–50(–70) μm wide; ascospores (9–)11–14(–17) × (3–)5–6 μm. 15. *Nectria peziza.*
       65. Mycelium conspicuous, white; perithecia globose or pyriform, not becoming cupulate, wall <20 μm wide; ascospores 6–16 × 3.5–7.5 μm.
        66. Ascospores (10.8–)11.3–14.2(–16) × (4–)4.5–6.2(–7.5) μm. 53. *Nectriopsis puiggarii.*
        66. Ascospores (6.3–)8.4–10.5(–11) × 3.5–4(–4.2) μm. 52. *Nectriopsis albofulta.*
      64. Ascospores smooth.
       67. Ascospores ellipsoidal with round ends, averaging <15 μm long.
        68. Mycelium not visible; perithecia globose, becoming cupulate, wall 30–50(–70) μm wide; ascospores (9–)11–14(–17) × (3–)5–6 μm. 15. *Nectria peziza.*
        68. Mycelium dense, white, perithecia globose to pyriform, becoming laterally pinched; ascospores, (7–)9.5–13(–14) × (3–)3.5–4(–5) μm. 50. *Nectriopsis byssotecta.*
       67. Ascospores fusiform with acute ends, (13.5–)14.8–17.3(–18) × 3.5–4.2(–4.5) μm. 51. *Nectriopsis oxyspora.*
  58. Perithecia seated in at most thin, white to tan mycelium.
   69. Perithecia with stiff, thick-walled setae.
    70. Setae <50 μm long; ascospores (12.5–)13.6–16.5(–17) × (2.5–)2.8–3.6(–4) μm, disarticulating at the septum; foliicolous. 4. *Trichonectria erythroxylifoliae.*
    70. Setae >50 μm long; ascospores disarticulating or not; not foliicolous.

71. Setae 75–120 μm long; ascospores (12–)14.5–18.5(–24.5) × (2.5–)3–4 μm, disarticulating at the septum; on *Diatrype stigma* and members of the Sphaeriaceae. 1. *Trichonectria rectipila.*

71. Hairs 50–80 μm long; ascospores (13–)13.5–16(–18) × (3–)3.5–4.5(–5) μm, not disarticulating at the septum; on members of the Sphaeriaceae. 2. *Trichonectria horrida.*

69. Perithecia with thin-walled hairs or glabrous.

72. Ascospores striate.

73. Perithecia seated in or on white mycelium; ascospores (6.3–)8.4–10.5(–11) × 3.5–4(–4.2) μm. 52. *Nectriopsis albofulta.*

73. Mycelium scant or not visible; ascospores 8–12 × 4–5 μm. 26. *Nectriopsis sororicola.*

72. Ascospores smooth or spinulose.

74. Ascospores (12–)13.7–17.2(–21) × (5–)5.7–7(–8) μm; on *Nectria.* 28. *Nectriopsis hyperbiota.*

74. Ascospores narrower; on various ascomycetes or not obviously fungicolous.

75. Perithecia with thin-walled hairs growing out from the papilla; on various ascomycetes and their anamorphs; ascospores (8–)9.5–13.5(–17) × (3.5–)4.2–6(–7) μm. 24. *Nectriopsis lasioderma.*

75. Perithecia lacking hairs; on various fungi or not obviously fungicolous; ascospores otherwise.

76. Ascospores 14.5–20 × (2.5–)3–5(–5.4) μm; on herbaceous debris and bamboo culms. 5. *Peristomialis parilis.*

76. Ascospores shorter and broader; on foliicolous or corticolous fungi.

77. Ascospores (11–)12–15(–17) × 4.5–5.5 μm, spinulose; on black fungus on a leaf. 58. *Nectriopsis cordiae.*

77. Ascospores (10–)11–12.5(–13) × 4.5–5.5(–6) μm, smooth; on members of the Diatrypaceae on wood. 56. *Nectriopsis flavella.*

Synoptic Key
Based on Substrate and Perithecial Characters

I. Occurring on superficial Ascomycetes

A. On *Nectria*
1. Ascospores averaging <10 × <3 μm
   7, 18, 19, 48, 57
2. Ascospores averaging <10 × >3 μm
   25, 26, 27, 48, 49, 52, 54, 57
3. Ascospores averaging >10 × <3 μm
   23, 28
4. Ascospores averaging >10 × >3 μm
   24, 25, 26, 28, 50, 51
5. Ascospores conspicuously striate
   26, 54, 52

B. On Xylariaceae
1. Ascospores <10 × <3 μm
   19, 55
2. Ascospores averaging >10 × <3 μm
   23
3. Ascospores averaging >10 × >3 μm
   12, 16, 17, 53
4. Ascospores conspicuously striate
   53

C. On Meliolaceae on leaves
1. Ascospores averaging <12 μm long
   33, 34
2. Ascospores averaging >12<16 μm long
   29, 30, 32
3. Ascospores averaging >16 μm long
   31
4. Ascospores striate
   31, 32

D. On inoperculate discomycetes
1. Ascospores averaging <14 × <3 μm
   3
2. Ascospores averaging <14 × >3 μm
   22
3. Ascospores averaging >15 × <3 μm
   21
4. Perithecia with thick-walled setae
   3

E. On operculate discomycete (*Sepultaria*)
   20

F. On black, superficial pyrenomycetes (excluding Xylariaceae, Meliolaceae)
  1. Ascospores averaging <10 × <3 μm
    13, 48, 55
  2. Ascospores averaging <10 × >3 μm
    13, 48
  3. Ascospores averaging >10 × <3 μm
    23
  4. Ascospores averaging >10 × >3 μm
    1, 2, 24
  5. Ascospores striate
    13
  6. Perithecia with thick-walled setae
    1, 2

G. On *Balansia* spp. on grass culms
  59, 60

H. On lichens
  18, 35

II. On Basidiomycetes
  15, 38

III. On Myxomycetes
  1. Ascospores averaging <6 μm long
    38, 39
  2. Ascospores averaging >6<8 μm long
    40, 41
  3. Ascospores averaging >8 μm long
    42, 43

IV. On immersed Ascomycetes or fungi or not obviously fungicolous (possibly on superficial Hyphomycetes or mycelium)

A. On bamboo
  1. Ascospores averaging <10 × <3 μm
    37, 44
  2. Ascospores averaging <10 × >3 μm
    36, 44
  3. Ascospores averaging >10 × <4 μm
    47
  4. Ascospores averaging >10 × >4 μm
    5
  5. Ascospores striate
    5

B. On bark and wood
  1. Ascospores averaging <7 × >3 μm
    38
  2. Ascospores averaging <10 × <3 μm
    18, 37, 46, 57
  3. Ascospores averaging <10 × >3 μm
    8, 46, 57
  4. Ascospores averaging >10<20 μm long
    15, 16, 56
  5. Ascospores averaging >20 μm long
    9, 11
  6. Ascospores phragmosporous
    9
  7. Ascospores dictyosporous
    11

C. On herbaceous debris (including palms, excluding bamboos)
  1. Ascospores averaging <10 × <3 μm
    37, 45
  2. Ascospores averaging >10<16 × <4 μm
    4, 37, 47
  3. Ascospores averaging >10<16 × >4 μm
    16, 57
  4. Ascospores averaging >16<25 μm long
    5, 14
  5. Ascospores averaging >25 μm long
    9, 10
  6. Ascospores phragmosporous, ≤5 septa
    10
  7. Ascospores phragmosporous, >5 septa
    9
  8. Perithecia with thick-walled setae
    4

Numerical List of Taxa

I. *Trichonectria*
  1. *T. rectipila*
  2. *T. horrida*
  3. *T. albidopilosa*
  4. *T. erythroxylifoliae*

II. *Peristomialis*
  5. *P. parilis*
  6. *P. paraparilis*
  7. *P. leucocarpa*
  8. *P. dentifera*
  9. *P. corynospora*
  10. *P. berkeleyi*
  11. *P. dictyospora*

III. *Hypocreopsis*
  12. *H. xylariicola*

IV. *Nectria*
  13. *N. brunneostriata*
  14. *N. tonduzi*
  15. *N. peziza*
  16. *N. suffulta*
  17. *N. discoidea*

V. *Nectriopsis*
  18. *N. mindoensis*
  19. *N. perpusilla*
  20. *N. sepultariae*
  21. *N. discophila*
  22. *N. discicola*
  23. *N. ostiolorum*
  24. *N. lasioderma*
  25. *N. lasiodermopsis*
  26. *N. sororicola*
  27. *N. sibicola*
  28. *N. hyperbiota*
  29. *N. cephalosporii*
  30. *N. sensitiva*
  31. *N. oidioides*
  32. *N. leucorrhodina*
  33. *N. guarapiensis*
  34. *N. pipericola*
  35. *N. porinicola*
  36. *N. hypocrellicola*
  37. *N. squamulosa*
  38. *N. oropensoides*
  39. *N. exigua*
  40. *N. candicans*
  41. *N. violacea*
  42. *N. sporangiicola*
  43. *N. hirsuta*

44. *N. microthecia*
45. *N. nanocarpa*
46. *N. septofusidiae*
47. *N. epimyces*
48. *N. cupulata*
49. *N. bactridioides*
50. *N. byssotecta*
51. *N. oxyspora*
52. *N. albofulta*
53. *N. puiggarii*
54. *N. epinectris*
55. *N. epimycota*
56. *N. flavella*
57. *N. lilliputia*
58. *N. cordiae*
59. *N. epichloe*
60. *N. macroepichloe*

Descriptions of Genera and Species

I. Trichonectria W. Kirschstein, Verh. Bot. Vereins Prov. Brandenberg **47:** 60. 1906 [1907]. Type: *T. aculeata* W. Kirschstein.

Perithecia superficial, nonstromatic, white to yellow, KOH+ or KOH−. Perithecial wall <20 μm wide, comprising small, nondescript cells; ostiolar canal periphysate. Thick-walled setae arising from the surface of the perithecial wall.

Included species: *Trichonectria albidopilosa, T. erythroxylifoliae, T. horrida, T. rectipila.*

The genus *Trichonectria* is characterized by having light colored perithecia from which arise thick-walled setae, and by phragmosporous ascospores (Rossman, 1983). At least one species, *T. hirta* (Bloxam) Petch is found on lichens. Thick-walled setae are rare among pyrenomycetes and especially among the Hypocreales. For this reason I accept *Trichonectria* and include in it these four species that have bicellular ascospores. The general aspect of the perithecium of *T. erythroxylifoliae* is unusual in the genus and it may be better placed elsewhere. At this time I cannot see any other place for it, however. Rossman (1983) considered *Trichonectria* to be related to *Paranectria* Sacc., a genus whose members are lichenicolous.

1. Trichonectria rectipila Samuels, Rogerson & Barr, sp. nov. (Fig. 1A–D).

Perithecia globosa, 100–170(–200) μm diam., non papillata, flavida. Setae rectae, simplices cylindricae, eseptatae, 75–150 × 7–10 μm. Ascosporae fusiformes, (12–)14.5–18.5(–24.5) × (2.5–)3–4 μm, bicellulares, in duas partes disarticulatae.

Status anamorphicus *Cephalosporiopsis*-similis. Holotypus. Pennsylvania (NY: *Samuels 86-560*).

Anamorph. *Cephalosporiopsis*-like.

Mycelium not evident. Perithecia solitary, superficial, nonstromatic, globose, 100–170(–200) μm diam., nonpapillate, collapsing by lateral pinching, pale yellow, not changing color in 3% KOH; with hairs white, stiff, erect, smooth, cylindrical or somewhat enlarged at the tip, aseptate, unbranched, 75–150 μm long × 7–10 μm wide, with lateral wall 3–4 μm thick and lumen ca. 2 μm wide, scattered over the upper half of the perithecial wall.

Cells at the surface of the perithecial wall 2–3 μm wide, forming *textura epidermoidea.* Perithecial wall ca. 10 μm wide, comprising a single region of flattened, ca. 5 μm long × 2–3 μm wide cells with walls ca. 0.5 μm thick. Ostiolar canal periphysate.

Asci clavate, (30–)35–50(–60) × 6–10 μm, apex simple; ≤8-spored, with frequent spore abortion; ascospores biseriate, completely filling each ascus or up to 25 μm of the ascal base devoid of ascospores. Ascospores fusiform, (12–)14.5–18.5(–24.5) × (2.5–)3–4 μm, equally 2-celled, rarely becoming 3-septate, often disarticulating at the median septum into two conical part spores; smooth, colorless.

Characteristics in Culture. Colonies grown 14 days at 20°C, diffuse daylight. CMD: 7 mm diam., flat, dense, opaque, aerial hyphae very short, pale salmon, eventually becoming yellow in the center of the colony. Conidia forming abundantly over the entire surface of the colony. Conidiophores arising from 1.5–2 μm wide hyphae on the surface of the agar and in the aerial mycelium, macronematous, mononematous, smooth, colorless, unbranched, 1-septate, subcylindrical, (17–)34–64(–76) μm long × 2–2.5 μm wide at base; monophialidic, phialides terminal, monoblastic, (17–)34–64(–76) μm long, tip slightly thickened, not flared, 1 μm wide. Conidia cylindrical, (12.6–)13–21.4(–30.6) × (1.8–)2.6–3.7 (–4.5) μm, base obtuse and lacking an obvious basal abscission scar, 0–1(–3)-septate, smooth, colorless, held in pale salmon slime.

Habitat. Stromata of *Diatrype stigma* Fr. and on Sphaeriaceae on *D. stigma.*

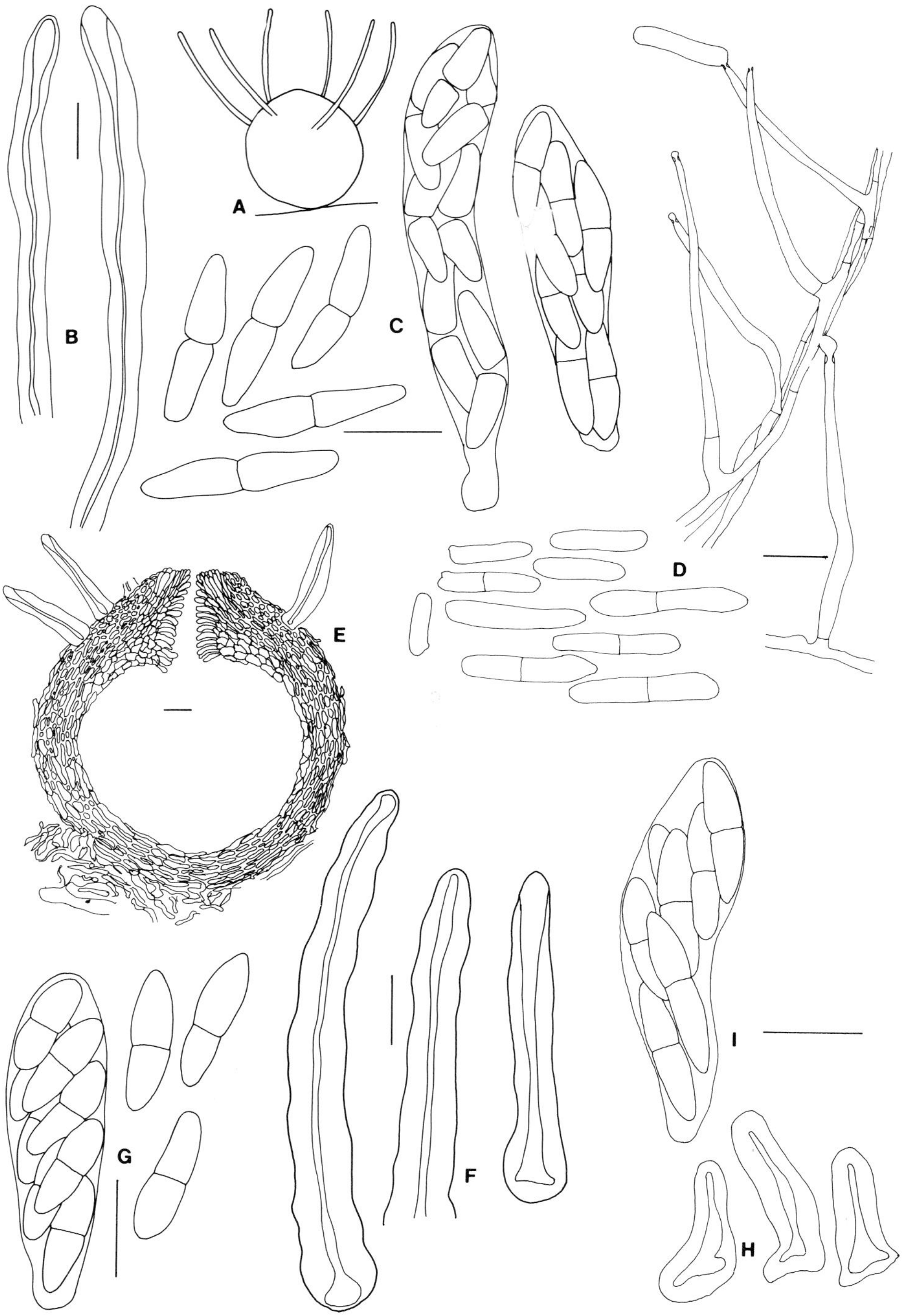

FIG. 1. A–D. *Trichonectria rectipila.* **A.** Perithecium (*Barr 6014*). **B.** Perithecial hairs (*Barr 6014*). **C.** Asci (*Barr 5034a*) and ascospores (*Barr 6014*). **D.** Conidiophores and conidia (*GJS 86-560*). E–G. *Trichonectria horrida.* **E.** Section through a mature perithecium. **F.** Perithecial hairs. **G.** Asci and ascospores. E–G drawn from *GJS 81-81.* H, I. *Trichonectria erythroxylifoliae.* **H.** Perithecial hairs. **I.** Ascus with ascospores. H, I drawn from the holotype. Lines = 10 μm.

Known Distribution. USA: Mai[illegible] Massachusetts, New York, Pennsylvania.

Etymology of the Specific Epithet. Refers to the stiff perithecial setae.

Holotype. USA. Pennsylvania: Chester County, Kennett Square, ½ mi E of Longwood Gardens, Hammorton Woods, on Sphaeriaceae on *Diatrype stigma, Rogerson,* 2[illegible] [illegible]986 (NY: *Samuels 86-560*).

Additional Specimens Examined. USA. MAINE: Franklin County, Carrabassett, Flagstaff Lake Rd., on old *Diatrype stigma* on *Betula* sp., *H. E. Bigelow & M. E. Barr 5834a,* 25 Aug 1971 (MASS, NY). MASSACHUSETTS: Franklin County, Conway State Forest, on *Fagus* wood with old pyrenos, *H. E. Bigelow & M. E. Barr 6014,* 26 Jul 1972 (MASS, NY). NEW YORK: Saratoga County, vic. Wilton, mixed woods, on *Diatrype stigma* on fallen branches, *S. J. Smith 45912 & Phelps,* 17 Oct 1970 (NY: *C. T. Rogerson 70-276*); Schuyler County, Arnot Forest of Cornell University, on inner bark of ?*Betula, Welch & Rogerson 2185 (part),* 2 Nov 1947 (NY).

2. Trichonectria horrida Samuels, sp. nov. (Fig. 1E–G).

Perithecia globosa vel late pyriformia, ca. 160 μm alta × 130–150 lata, flavida. Setae rectae, simplices, eseptatae, 50–80 μm longae, basi 9–10 μm et apicem versus 5.5–6.5 μm latae. Ascosporae fusiformes vel ellipsoideae, (13–)13.5–16(–18) × (3–)3.5–4.5(–5) μm; bicellulares.

Status anamorphicus ignota.
Holotypus. New Zealand (PDD: *Samuels 81-81*).

Mycelium not apparent. Perithecia solitary, superficial, nonstromatic, globose to broadly pyriform, ca. 160 μm high × 130–150 μm wide, nonpapillate, collapsing by lateral pinching, at first pale yellow but becoming amber colored, not changing color in 3% KOH; with white, stiff, erect, smooth, subacute, unbranched, aseptate setae, 50–80 μm long × 9–10 μm wide basally × 5.5–6.5 μm wide apically, lateral wall 3–4 μm thick, wall at tip thinner, lumen 1–2 μm wide, arising from the upper third of the perithecial wall.

Perithecial wall covered with ca. 3 μm wide hyphae with walls 0.5 μm thick. Perithecial wall 15–20 μm wide, consisting of a single region of cells; cells ± elliptical in section, 5–10 × 2–3 μm, walls 0.5 μm thick; cells at perithecial apex elliptical, ca. 3 × 2 μm; ostiolar canal periphysate.

Asci clavate, 35–50 × 10–13(–15) μm, apex broadly rounded, simple; 8-spored, ascospores arranged without order and completely filling each ascus. Ascospores fusiform to ellipsoidal, (13–)13.5–16(–18) × (3–)3.5–4.5(–5) μm, equally 2-celled or apical cell somewhat longer and broader than the basal cell, not constricted at the septum, smooth, colorless.

Characteristics in Culture. Colonies grown three weeks at 15–18°C, diffused daylight on CMD barely growing away from the original inoculum, salmon-colored, sterile.

Habitat. On perithecia and associated mycelium of a sphaeriaceous pyrenomycete, possibly *Chaetosphaeria* sp.

Known Distribution. New Zealand: Auckland, known only from the type.

Etymology of the Specific Epithet. Refers to the setose perithecia.

Holotype. New Zealand. North Island. Auckland: Waitemata City, Waitakere Ranges, Swanson, on pyrenomycete (?*Chaetosphaeria*), *Samuels 81-81, Johnston & Horak,* 7 May 1981 (PDD).

Note. Unfortunately, cultures derived from ascospores of this species were lost before conidia were formed.

3. Trichonectria albidopilosa (Rogerson & Samuels) Samuels, comb. nov.

≡*Nectria albidopilosa* Rogerson & Samuels, Mycologia **77:** 781. 15 Oct 1985.

Anamorph. ?*Acremonium* sp.

Mycelium thin, white. Perithecia globose to broadly pyriform, 145–175 × 135–175 μm, nonpapillate, pale yellow; with stiff, erect, smooth, white, unbranched, 30–60 μm long × 7–10 μm wide hairs arising from the surface of the perithecial wall. Perithecial wall ca. 10 μm wide. Asci clavate, (32–)34–45(–47) × (5–)6–8 μm, apex simple. Ascospores narrowly ellipsoidal to fusiform, (10–)11.6–13.5(–14.5) × 2–3 μm, equally 2-celled, colorless, smooth.

Habitat. On the hymenium of inoperculate discomycetes.

Note. For illustrations and a complete discussion see Rogerson and Samuels (1985).

4. Trichonectria erythroxylifoliae (Viégas) Samuels, comb. nov. (Fig. 1H, I).

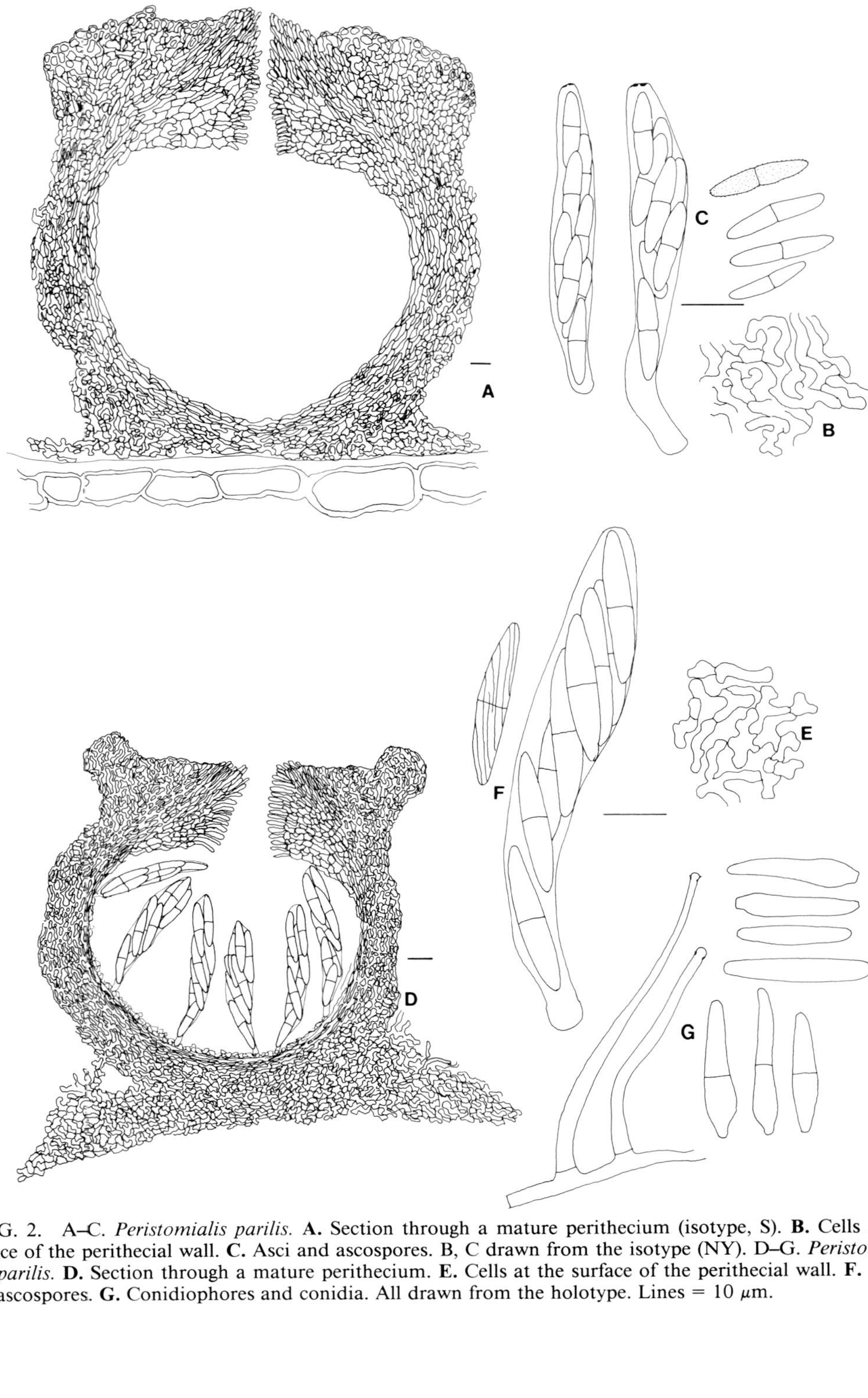

FIG. 2. A–C. *Peristomialis parilis.* **A.** Section through a mature perithecium (isotype, S). **B.** Cells at the surface of the perithecial wall. **C.** Asci and ascospores. B, C drawn from the isotype (NY). D–G. *Peristomialis paraparilis.* **D.** Section through a mature perithecium. **E.** Cells at the surface of the perithecial wall. **F.** Ascus and ascospores. **G.** Conidiophores and conidia. All drawn from the holotype. Lines = 10 μm.

≡*Nectria erythroxylifoliae* Viégas, Bragantia, **3:** 52. 1943.

Anamorph. None known.

Mycelium tan, spreading, thin; hyphae 3–4 μm wide, branched, septate. Perithecia scattered, solitary to gregarious, nonstromatic, easily removed from the substrate, globose, 130–170 (–190) μm diam., nonpapillate, becoming cupulate when dry; with scattered setae arising from cells of the upper half of the perithecial wall; setae stiff, erect, unbranched, aseptate, acute, 25–35 μm long × ca. 10 μm wide at base, wall 2–3 μm.

Cells at surface of perithecial wall hyphal, cells ca. 3 μm wide, tightly compacted. Perithecial wall 10–15 μm wide, comprising a single region of ± fusoid, ca. 7 × 2 μm cells with walls ca. 1 μm thick; ostiolar canal periphysate.

Asci clavate, 35–40 × 9–11 μm, apex simple; 8-spored, ascospores multiseriate, completely filling each ascus. Ascospores fusiform, (12.5–) 13.6–16.5(–17) × (2.5–)2.8–3.6(–4) μm, equally 2-celled, often deeply constricted at the septum and tending to disarticulate while still in the ascus, smooth, colorless.

Habitat. On leaves of *Erythroxylum suberosum* St. Hil., not obviously fungicolous.

Known Distribution. Brazil, known only from type.

Holotype. Brazil. Minas Gerais: Cerrado, Lagoa Santa, on leaves of *Erythroxylum suberosum*, *Viégas, Barreto & Krug,* 8 Apr 1936 (IACM).

Notes. In spite of the thick-walled perithecial setae, it is unlikely that *T. erythroxylifoliae* is closely related to other species of *Trichonectria.*

II. **Peristomialis** (Phillips) Boudier, Hist. Class. Discom. Europe 116. 1907.

≡*Mollisia* [subg.] *Peristomialis* Phillips, Man. Brit. Discom. 201. 1887. Lectotype: *P. berkeleyi* Boudier, *vide* Rogerson, Mycologia **62:** 897. 1970.

Perithecia superficial, nonstromatic, white to yellow, KOH+ or KOH−. Perithecial wall <20 μm wide, perithecial apex discoidal, the disc formed of intertwined hyphae that may grow outwardly in triangular fascicles to form a distinct crown of hyphal hairs.

Included species: *Peristomialis corynospora, P. berkeleyi, P. dentifera, P. dictyospora, P. leucocarpa, P. paraparilis, P. parilis.*

Peristomialis was originally described with six species; Rogerson (1970) designated *P. berkeleyi* as the lectotype. I (Samuels, 1976b) have previously maintained *Peristomialis* as a synonym of *Nectria* but *P. berkeleyi* shares few characters with *N. cinnabarina* (Tode: Fries) Fries, the type species of *Nectria.* Species of *Peristomialis* are easily recognized by the anatomy of the perithecial wall and by the apical disc, especially those species the perithecia of which have triangular hyphal hairs radiating from the perithecial apex. These fungi are most often found on decaying herbaceous debris where the true substrate may be other fungi; *P. parilis* is unequivocally associated with immersed ascomycetes on leaves of *Podocarpus* sp.

Perithecia of some species of *Nectria* (e.g., *N. suffulta* Berk. & Curtis, *N. sylvana* Mouton) resemble those of species of *Peristomialis* in having triangular, fasciculate hyphal hairs arising from the surface of the perithecial wall. However, the perithecial wall anatomy of these species is very different from the wall anatomy of perithecia of *Peristomialis* species.

Anamorphs, known for *Peristomialis berkeleyi, P. corynospora, P. dentifera, P. leucocarpa,* and *P. paraparilis* are either species of *Acremonium* or *Cephalosporiopsis* (herewith and Samuels, 1976b).

5. **Peristomialis parilis** (H. Sydow) Samuels, comb. nov. (Fig. 2A–C).

≡*Nectria parilis* H. Sydow, Ann. Mycol. **28:** 121. 20 May 1930.

Anamorph. None known.

Mycelium not evident. Perithecia solitary or in groups of a few, superficial, nonstromatic, globose, 180–340 μm diam.; with a flat apex 110–150 μm diam.; and sometimes with short, fasciculate, triangular hairs or with wart-like processes, each comprising intertwined hyphae, growing out from the apical disc; smooth, not collapsing when dry, pale yellow to white, not changing color in 3% KOH.

Surface of perithecial wall *textura epidermoidea.* Perithecial wall 20–25 μm wide, composed of intertwined hyphae, cells in section elliptical, ca. 5 × 3 μm, walls ca. 1 μm thick; perithecial apex of intertwined hyphae; ostiolar canal periphysate.

Asci clavate, (40–)45–65(–75) × 8–11(–13) μm, apex with an obscure ring, sometimes appearing simple; 8-spored, ascospores biseriate, completely filling each ascus. Ascospores fusiform, 14.5–20 × (2.5–)3–5(–5.4) μm, equally 2-celled, slightly constricted, markedly spinulose, colorless.

Habitat. On herbaceous debris including leaves of *Podocarpus* and bamboo culms.

Isotypes. Venezuela. Aragua: prope Colonia Tovar, in foliis *Podocarpi coriacei* Rich., saepe in consortio aliorum fungorum, *H. Sydow* (Fungi exotici exsiccati 838), 19.1.1928 (NY, S).

Additional Specimens Examined. FRENCH GUIANA. Saül, 200 m, on decaying culm of *Guadua* sp., *Samuels 3777,* 3–16 Feb 1986 (NY); ca. 150 km S of St. Laurent du Maroni, Citron, Paul Isnard area, Mt. Lucifer, on black fungus on dead bamboo culm, *Samuels 4158,* 10–14 Mar 1986 (CAY, BPI, PDD, NY). USA. CALIFORNIA: On bark of *Eucalyptus globulus* La Bill., *Dr. Harkness 2359,* date of collection not given (K, as type of *Nectria coronata*). VENEZUELA. Data as holotype, *H. Sydow* (NY: Fungi exotici 759; issued as *Physalacria tenera* H. Sydow. A probable isotype of *Nectria parilis*).

Notes. *Peristomialis parilis* is not always obviously fungicolous but perithecia of one of the isotypes (NY: Fungi exotici 838) clearly grow out of the ascomata of a black, immersed ascomycete. Perithecia of the specimens from French Guiana are associated with several fungi including the hyphomycete *Acrodictys bambusicola* M. B. Ellis. Ascospores (35) of the isotype collections measure (13.5–)14.5–18.2(–21) × (2.5–)2.7–3.5 μm and are shorter and narrower than spores found in the collections from French Guiana [n = 32, (15.3–)16.8–21(–27) × 4.5–5.5 μm] and California [n = 22, (15–)15.7–19.4(–20) × 3.9–4.2(–4.5) μm].

The specimen from California cited above was deposited as *Nectria coronata* without any indication of the authority of the species or of the identifier. The perithecia are clearly distinct from *N. coronata* Penzig & Saccardo; no reference to another use of the name *N. coronata* has been found, thus the name used in this connection is assumed to be a herbarium name.

Peristomialis parilis is most closely related to *P. paraparilis,* differing in having shorter, striate ascospores.

6. **Peristomialis paraparilis** Samuels, sp. nov. (Fig. 2D–G).

Peristomialidis parili (H. Sydow) Samuels similis sed ascosporis (19–)21.5–24.5(–25.5) × 4–5 μm.

Status anamorphicus *Acremonium* sp.
Holotypus. French Guiana (NY: *Samuels 4159*).
Anamorph. *Acremonium* sp.

Mycelium not evident. Perithecia solitary to gregarious, superficial, nonstromatic, anchored to substrate by densely compacted and intertwined hyphae, easily removed from the substrate, globose to subglobose, 155–185 × 155–175 μm; with a flat apical disc 100–130 μm diam.; smooth, not collapsing when dry, pale yellow, nearly white, not changing color in 3% KOH; perithecial contents pale orange.

Cells at the surface of the perithecial wall forming *textura epidermoidea.* Perithecial wall 20–25 μm wide, comprising a single region of intertwined hyphae that, at the apex, form an apical disc; ostiolar opening formed by periphyses that grow through the hyphal layer of the apical disc, ostiolar canal periphysate.

Asci clavate, (63–)65–75(–80) × (12.5–)13.5–17(–18) × 6.5–7.5(–8) μm, apex simple; 8-spored, ascospores biseriate, completely filling each ascus. Ascospores fusiform, (19.5–)21.5–24.5(–25.5) × 4–5 μm, equally 2-celled, not constricted at the septum, becoming coarsely striate, colorless.

Habitat. On black mycelium on bamboo culm.

Known Distribution. French Guiana, known only from the type.

Etymology of the Specific Epithet. Refers to the similarity of the perithecia to those of *P. parilis.*

Holotype. French Guiana. Ca. 150 km S of St. Laurent du Maroni, Citron, Paul Isnard area, Mt. Lucifer, on black fungus on dead bamboo culm, *Samuels 4159,* 7–14 Mar 1986 (NY; isotype CAY).

Notes. *Peristomialis paraparilis* falls neatly between *P. parilis* and *P. berkeleyi* (see Samuels, 1976b) in having perithecia of the former and ascospores of the latter.

The type collection of *Peristomialis paraparilis* is associated with *Acrodictys bambusicola* and other dark hyphal and dark-spored fungi.

7. **Peristomialis leucocarpa** Samuels, sp. nov. (Fig. 3A–C).

Peristomialidis dentiferae Samuels similis sed ascosporis (7.5–)8.5–10(–11) × 2–2.7(–3) μm.

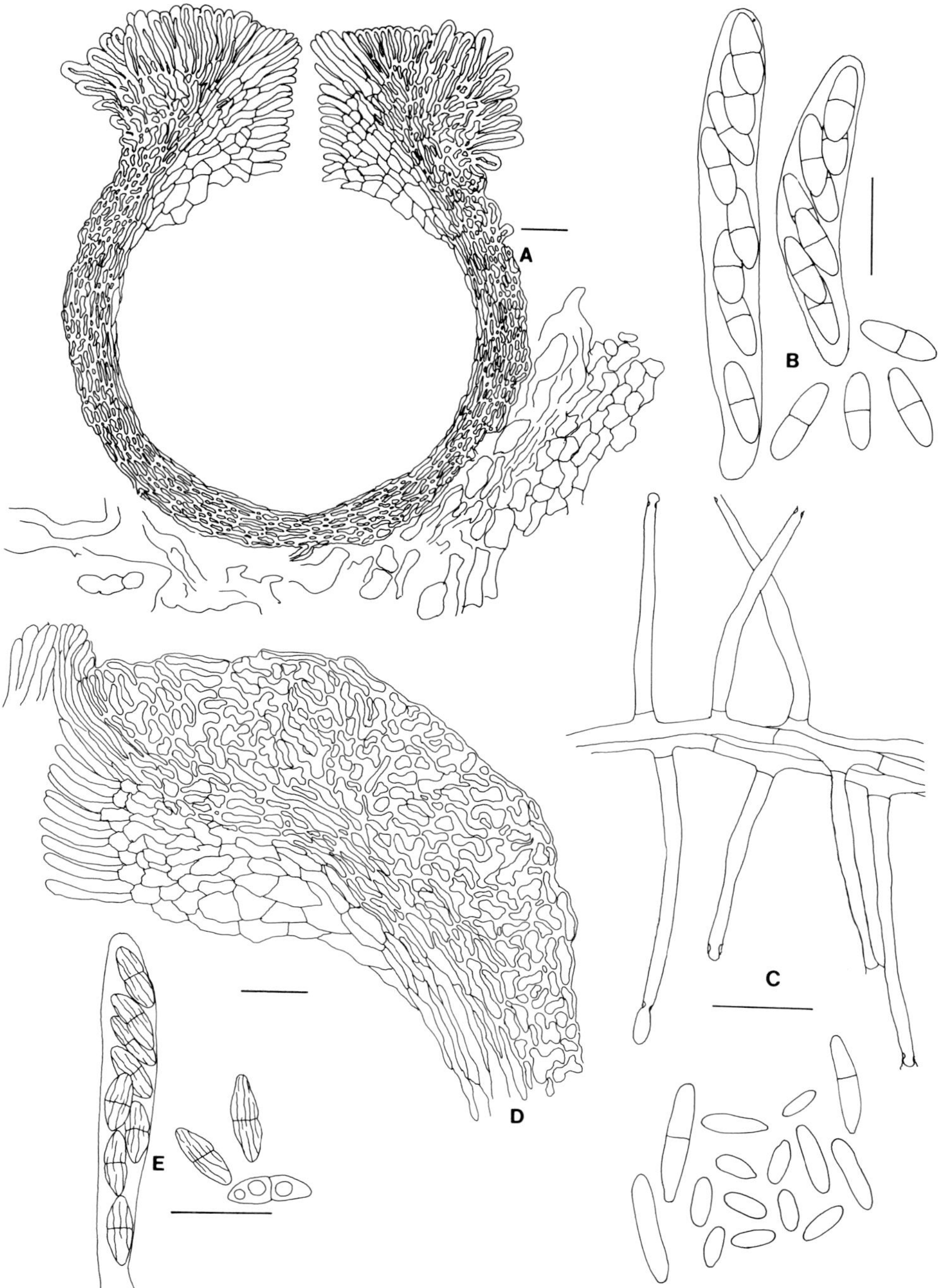

FIG. 3. A–C. *Peristomialis leucocarpa.* **A.** Section through a mature perithecium. **B.** Asci and ascospores. **C.** Conidiophores and conidia. All drawn from the holotype. D, E. *Nectria brunneostriata.* **D.** Portion of the upper perithecial wall and ostiolar canal of a mature perithecium (*Samuels 2410*). **E.** Asci and ascospores (*Samuels 2392*). Lines = 10 μm.

Status anamorphicus *Acremonium* sp.
Holotypus. Jamaica (NY: *CUP-MJ 830*).
Anamorph. *Acremonium* sp.

Mycelium not apparent. Perithecia solitary, scattered, superficial, nonstromatic, globose to subglobose, (105–)110–150(–175) μm high × (80–)95–144(–160) μm wide; with a flat apical disc 90–100 μm diam; collapsing by lateral pinching when dry; white, not changing color in 3% KOH.

Perithecial wall covered with hyphae. Perithecial wall ca. 10 μm wide, comprising a single region of intertwined hyphae, cells appearing ± elongated, 2–5 μm long × ca. 1 μm wide, walls <0.5 μm thick; apical disc formed of intertwined hyphae, hyphae of the disc sometimes growing outward to form short hairs; ostiolar canal formed by periphyses pushing through the hyphae of the apical disc.

Asci clavate, (30–)33–48(–50) × (5–)6–8 μm, apex simple; 8-spored, ascospores completely to partially biseriate, completely filling each ascus. Ascospores ellipsoidal to oblong, (7.5–)8.5–10 (–11) × 2–2.7(–3) μm, equally 2-celled, not constricted at the septum, smooth, colorless.

Characteristics in Culture. Colonies grown two weeks at 20°C, 12 hours dark/12 hours near ultraviolet + cool white fluorescent light. CMD: 2 cm diam., felty, opaque, aerial hyphae short; pale yellow pink, near white. Conidia forming abundantly on all media; conidiophores 30–35(–45) μm long × 2 μm wide at the base, arising from 1.5–2 μm wide hyphae in aerial mycelium, from poorly developed, hyphal, rope-like strands, and from the surface of the agar, macronematous, mononematous, straight, smooth; with a single, basal, septum; monophialidic, phialides 30–35 (–45) μm long, tip 1 μm wide, slightly thickened, not flared. Conidia oblong to ellipsoidal, (4–)4.5–10(–16) × (1.5–)1.7–2(–3.5) μm, 0(–1)-septate, with or without a protuberant, flat, basal abscission scar, produced in basipetal succession and held in colorless drops of liquid.

Habitat. On old perithecia of *Nectria jungneri* Hennings and on surrounding tissue of pod of *Theobroma cacao* L.

Known Distribution. Jamaica, known only from type.

Etymology of the Specific Epithet. Refers to the white perithecia.

Holotype. Jamaica. St. Thomas Parish: Along Sulphur River above Bath Fountain Hotel, 500 ft, on *Nectria jungneri* and pod of *Theobroma cacao, Korf, Dixon, Dumont, Erb, Pfister, Reynolds, Rossman & Samuels,* 13 Jan 1971 (NY: *CUP-MJ 830*).

Note. *Peristomialis leucocarpa* differs from *P. dentifera* (Samuels, 1976a) in having larger ascospores.

8. Peristomialis dentifera (Samuels) Samuels, comb. nov.

≡*Nectria dentifera* Samuels, New Zealand J. Bot. **14:** 253. 1976.

Anamorph. *Cephalosporiopsis* sp.

Mycelium not evident. Perithecia superficial, nonstromatic, globose with a flat apex, 200–250 μm diam., yellow to orange, with short, tooth-like processes composed of triangular fascicles of hyphae arising from around the perithecial apex. Asci cylindrical, 45–55 × 5–6 μm, apex with a ring. Ascospores ellipsoidal, 6–8 × 3–4 μm, 1-septate, finely spinulose, colorless.

Habitat. On bark and decorticated wood.

Note. For illustrations and a complete description see Samuels (1976b, as *Nectria dentifera*).

9. Peristomialis corynospora (Samuels) Samuels, comb. nov.

≡*Nectria corynospora* Samuels, New Zealand J. Bot. **16:** 78. 1978.

Anamorph. *Cephalosporiopsis* sp.

Mycelium not evident, or appearing as a ring of white hyphae around each perithecial base. Perithecia superficial, nonstromatic, globose to pyriform with a flat apex, white, 150–250 μm diam., with an ostiolar disc of intertwined hyphae. Asci clavate, (50–)60–80(–87) × (6–)8–11 μm, apex with an obscure ring, often appearing simple. Ascospores clavate to fusiform, 27–42 × 3–4.5 μm, 5–9(–13)-septate, smooth, colorless.

Habitat. On woody and herbaceous debris.

Notes. For illustrations and complete descriptions see Samuels (1978, as *Nectria corynospora*) and Rossman (1983, as *Nectria opalina*). Rossman (1983) gave *Nectria corynospora* as a synonym of *N. opalina.* However, perithecia of *N. opalina* were originally described as being half immersed in the substrate, and with a papilla (Crouan & Crouan, 1867). Ms. Rosalind Lowen (NY) has reexamined the type collection of *N.*

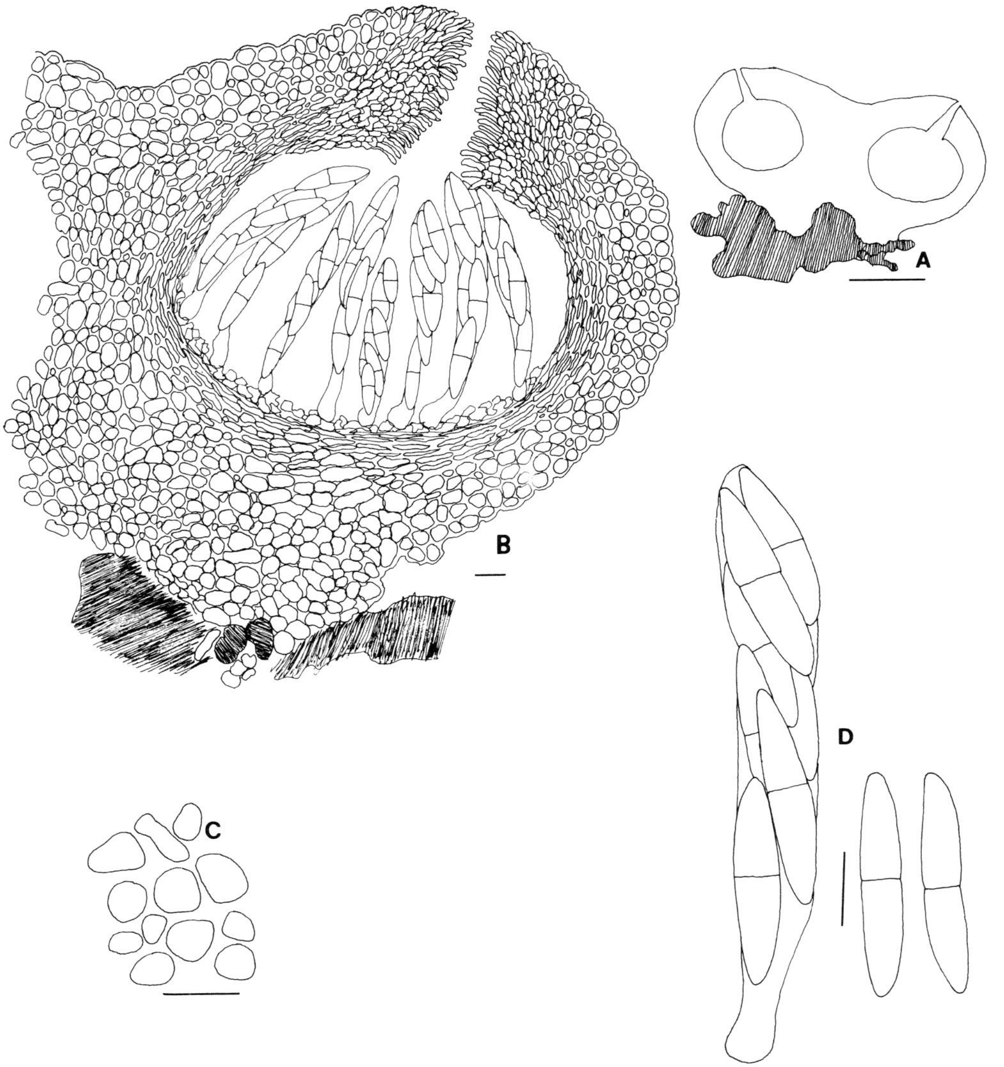

FIG. 4. *Hypocreopsis xylariicola.* **A.** Diagram of two perithecia immersed in a stroma. **B.** Section through a mature perithecium. **C.** Cells at the surface of the stroma. **D.** Asci and ascospores. All drawn from the holotype. Lines A = 200 μm, B–D = 10 μm.

opalina (CO), finding an immersed ascomycete with light-colored perithecia and phragmosporous ascospores; the ascal apex was thickened, however, typical of the Clavicipitales. Because perithecia of *Peristomialis corynospora* are entirely superficial, and because its asci lack apical thickening, it is unlikely that *P. corynospora* and *N. opalina* are conspecific. Rossman's (1983) description agrees in all details with *P. corynospora,* thus suggesting that at least one perithecium of the species was indeed present with perithecia of the type of *N. opalina.*

10. Peristomialis berkeleyi Boudier, Histoire et Classification des Discomycetes d'Europe. p. 116. 1907.

≡*Peziza peristomialis* Berkeley & Broome, Ann. Mag. Nat. Hist. Ser. 3, **18:** 126. Aug 1866.
≡*Nectria peristomialis* (Berk. & Broome) Samuels, New Zealand J. Bot. **14:** 251. 1976.

≡*Nectria tympaniformis* Petrak, Sydowia **4:** 514. 30 Sep 1950.

Anamorph. *Acremonium* sp.

Mycelium not evident, or white, scant, and spreading. Perithecia superficial, nonstromatic, globose with a flat apex, 200–370 μm diam., colorless with orange contents, with conspicuous, radiating, fasciculate hyphal hairs. Asci clavate, (52–)70–85(–105) × (8–)9–13(–15) μm, apex simple. Ascospores fusiform, (18–)25–32(–37) × (3–)4–6(–7) μm, 1–2(–5)-septate, striate, colorless.

Habitat. On herbaceous debris.

Note. For illustrations, complete descriptions, and additional synonyms see Samuels (1976b, as *Nectria peristomialis*) and Rossman (1983, as *Nectria peristomialis*).

11. Peristomialis dictyospora (Rossman) Samuels, comb. nov.

≡*Nectria dictyospora* Rossman, Mycol. Pap. **150:** 69. Feb 1983.

Anamorph. None known.

Mycelium not evident. Perithecia superficial, nonstromatic, globose to subglobose with a flat apex, 230–375 μm high × 260–350 μm wide, with conspicuous, radiating, fasciculate hyphal hairs. Asci clavate, 70–100 × 20–25 μm, apex simple. Ascospores clavate to fusiform, 48–97 × 10–16 μm; with 7–13 transverse septa and one, irregular, longitudinal septum; smooth, colorless.

Habitat. Forming on wood.

Notes. *Peristomialis dictyospora* is morphologically similar to *P. berkeleyi*. For illustrations and a complete description see Rossman (1983, as *Nectria dictyospora*).

III. Hypocreopsis P. A. Karsten, Mycol. Fenn. **2:** 251. 1873. Type: *H. riccioidea* (Bolton: Fr.) P. A. Karsten (*Sphaeria riccioidea* Bolt.: Fr.).

Perithecia completely to partially immersed in a stroma, reddish brown or yellow, KOH+ or KOH−. Perithecial wall thin or thick and variously constructed. Setae and hyphal hairs lacking. Ascospores bicellular, colorless.

Included species: *Hypocreopsis xylariicola.*

Hypocreopsis was originally described with the single species, *H. riccioidea* (≡*H. lichenoides* (Tode: Fr.) Seaver). Müller and Arx (1962) limited the genus to six species. Doi (1967) removed *H. phyllostachydis* (H. & P. Sydow) Miyake & Hara to *Mycocitrus* Möller, and Kobayashi and Doi (1969) excluded *H. hypoxyloides* Speg (=*Phaeocreopsis hypoxyloides* (Speg.) Sacc. & P. Sydow) because of its bitunicate asci. Doi (1977) later added two more species, and Rifai (1969) added one. *Hypocreopsis macrostoma* (Berk. & Curt.) Müller is *Nectria sensu lato,* thus *Hypocreopsis* today includes eight species with the addition of *H. xylariicola.*

Three species of *Hypocreopsis* are known to be fungicolous, *H. lichenoides, H. episphaeria* (Pat.) Müller, and *H. xylariicola.* Because stromata of *H. lichenoides* and *H. episphaeria* are brown to reddish-brown, these species are not included in the present work. *Hypocreopsis lichenoides* has been found on *Hymenochaete* species (Cauchon & Ouellette, 1964) and *H. episphaeria* occurs on *Amphisphaeria discoidea* Pat.; *H. xylariicola* occurs on *Xylaria* sp.

The characters used to delimit *Hypocreopsis* are generalized, thus the genus is not natural. It includes species with KOH+ and others with KOH− stromata. Stromal anatomy is diverse, and stromal form varies from the grotesque, radiately lobed aspect presented by the type species to discrete and minute stromata of *H. tingomariensis* Doi and *H. xylariicola.* Anamorphs, known for only two species, are also diverse being sporodochial (*Stromatocrea* W. B. Cooke) in *H. lichenoides* (Petrak, 1952; Cauchon & Ouellette, 1964; Strid, 1967), or mononematous (*Paecilomyces*-like) in *H. subiculoides* Doi (Doi, 1977). *Hypocreopsis* is clearly a candidate for a revision but that is beyond the scope of the current work; thus I retain the broad concept of the genus *pro tem.*

12. Hypocreopsis xylariicola Samuels, sp. nov. (Fig. 4).

Mycelium extus non visibile. Perithecia 175–225 μm diam., complete in stromate immersa, aut solitaria, tunc non-stromatica, aut aggregata tunc ± tuberculato botryosa utriusque ambitu saepissime distinguibili. Stroma 50–60 μm latum et e regionibus cellularibus duabus constans, regione exteriore 25–30 μm crassa, cellulis ± angulatis, ca. 5 μm diam., muris plerumque ca. 1 μm crassis sed cellularum exteriorum juxta superficiem 3 μm crassis; regione interiore ca. 15 μm crassae, cellulis ellipticis, 10–12 × 2–2.5 μm, muris 1–1.5 μm crassis. Perithecii ostiolum e hyphis angustis cum regione interiore continuis per regionem externam productis constans, canali ostiolari periphysato. Stromatis cellulae interperitheciales et illae baseos stromatis 7–10 μm diam., ± angulares, muris ca. 0.5 μm crassis. Asci clavati, (58–)67–86(–90) × (5.5–)8.8–13.5 μm, apice simplices, 8-spori. Ascosporae fusiformes, (26–)27.8–30.6 × 4.5–5.7(–6) μm, laeves, hyalini, aequaliter 2-cellulares, non constricti sed maturitatem versus 2-septifer.

Hab. in *Xylaria* sp.
Status anamorphicus ignotis.
Holotypus. Guyana (NY: *Samuels 5045*).
Anamorph. None known.

Mycelium not visible, completely filling old perithecia of *Xylaria* sp. and growing out of ostiolar opening to form a stroma. Perithecia completely immersed in a stroma with outlines of individual perithecia visible, with one to a few perithecia in each stroma, sometimes appearing to be nonstromatic, perithecial aggregates ± tuberculate-botryose with outlines of individual perithecia usually apparent and globose, 175–225 μm diam., nonpapillate, smooth, not collapsing when dry, yellow, not changing color in 3% KOH.

Cells at surface of stroma angular in outline, 8–10 μm in greatest dimension, walls 1–1.5 μm thick. Stroma 50–60 μm wide, comprising two intergrading regions. Outer region 25–30 μm wide, cells ± angular in outline, ca. 5 μm in greatest dimension with walls ca. 1 μm thick but with external walls of cells at exterior of the stroma ca. 3 μm thick; inner region ca. 15 μm wide and forming the perithecial wall, cells elliptic in outline, 10–12 × 2–2.5 μm, walls 1–1.5 μm thick but becoming progressively thinner toward the interior of the locule. Ostiolar opening formed of narrow hyphal elements continuous with the interior region of the perithecial wall and protruding through the outer region, intergrading with the outer region of the stroma and with the periphyses. Cells of the stroma between and below perithecia angular, 7–10 μm in greatest dimension, walls ca. 0.5 μm thick.

Asci clavate, (58–)67–86(–90) × (5.5–)8.8–13.5 μm, apex simple; 8-spored, ascospores completely filling each ascus. Ascospores fusiform, (26–)27.8–30.6 × 4.5–5.7(–6) μm, equally 2-celled but with two septa forming with age, not constricted at the septa, smooth, colorless.

Habitat. On stromata of *Xylaria* sp.
Known Distribution. Guyana, known only from type collection.
Etymology of the Specific Epithet. Refers to the substrate.

Holotype. Guyana. Cuyuni-Mazaruni Region, VII; Mazaruni Sub-Region, VII-2. Along the Koatse R. ca. 2 km E of Pong R, ca. 5 hr walk S of Chinoweing village, ca. 05°28′N, 60°04′W, 600–650 m, on *Xylaria* sp., *Samuels 5045, Pipoly & Gharbarran,* 28 Feb 1987 (NY; isotype BRG).

Notes. This species is distinguished by its large spores, the largest known in the genus, and by the cells at the surface of the stromal wall that have distinct outlines and unevenly thickened walls (when seen in section).

IV. Nectria (Fries) Fries, Summa Veg. Scand. 387. 1849.

≡*Hypocrea* [sect.] *Nectria* Fries, Syst. Orbis Veget. 105. 1825. Lectotype: *N. cinnabarina* (Tode: Fries) Fries (*Sphaeria cinnabarina* Tode: Fries), *vide* Clements & Shear, Gen. Fungi 282. 1931.

Perithecia nonstromatic or seated on a conspicuous to nonconspicuous basal stroma, superficial, red (KOH+) or orange (KOH−). Perithecial wall >20 μm wide, cells in section angular, globose, or lacking a definite outline but wall not formed of conspicuously intertwined hyphae. Ascospores ellipsoidal, less frequently oblong or fusiform, nonapiculate, one to multiseptate, septa transverse, or transverse and longitudinal.

Anamorphs phialidic; pycnidial or hyphomycetous and mononematous, or sporodochial.

Five hyperparasitic species with yellow perithecia are retained *pro tem.* in *Nectria.* These species are not closely related to *N. cinnabarina,* the type species of *Nectria,* but their ultimate disposition must await more complete study of the nectrioid fungi.

Included species: *Nectria brunneostriata, N. discoidea, N. peziza, N. suffulta, N. tonduzi.*

Nectria peziza and *N. suffulta* are undoubtedly related to each other; *N. discoidea* may be related to these species. *Nectria tonduzi* is apparently related to *N. ochroleuca* (Schw.) Berk. I do not know of any close relatives for *N. brunneostriata.*

Nectria peziza and *N. suffulta* are not often found growing on other fungi whereas the remaining three species are found only on other fungi. *Nectria ochroleuca* and its *Clonostachys* anamorph (*Gliocladium roseum* Bain.) are often cited as agents of biological control of fungi, and *G. roseum* is a common soil fungus (see references in Domsch et al., 1980). I have not found perithecia of this species growing directly on other fungi.

13. Nectria brunneostriata Samuels, sp. nov. (Fig. 3D, E).

Perithecia pyriformia vel subglobosa, 150–250 × 100–200 μm, non papillata, laevia, flavida; perithecii paries 20–30 μm latus, ex hyphis intertextis compositus. Asci clavati, (25–)33–45 × (4.5–)4.8–6.8(–9) μm, apice simplices. Ascosporae fusiformes, (6.3–)6.7–8.2(–9) × 2.5–3.2 (–3.6) μm, 1-septatae, striatae, flavido-brunneae.

Status anamorphicus ignota.
Holotypus. Indonesia (BO: *Samuels 2410*).
Anamorph. None known.

Mycelium not evident. Perithecia cespitose in groups of few to many, superficial, seated on a thin stroma of compacted hyphae or on a pseudoparenchymatous stroma, difficult to remove from the substrate, pyriform to subglobose, 150–250 μm high × 100–200 μm wide, nonpapillate or with an acute papilla, smooth, collapsing by lateral pinching or not collapsing when dry, yellow to pale orange, not changing color in 3% KOH.

Cells at the surface of the perithecial wall *textura epidermoidea* with thickened walls. Perithecial wall 25–30 μm wide, of a single region of indistinct cells. Perithecial apex of nondescript small cells, ostiolar canal periphysate.

Asci clavate, (25–)33–45 × (4.5–)4.8–6.8(–9) μm, apex simple; 8-spored, ascospores partially to completely biseriate, completely filling each ascus. Ascospores fusiform, (6.3–)6.7–8.2(–9) × 2.5–3.2(–3.6) μm, with subacute ends, equally 2-celled, slightly constricted at the septum, coarsely to faintly striate, transparent yellow-brown.

Habitat. Perithecia of pyrenomycetes.

Known Distribution. Indonesia (North Sulawesi), known only from the type locality.

Etymology of the Specific Epithet. Refers to the brown, striate ascospores.

Holotype. Indonesia. North Sulawesi: Dumoga-Bone National Park, Gn. Muajat, Danau Alia; 00°45′N, 124°25′E, 1400 m, on black pyrenomycete (*Samuels 2398*) on flowering stalk of Zingiberaceae, *Samuels 2410,* 26 Oct 1985 (BO; isotype NY).

Additional Collections Examined. Data as holotype, *Samuels 2392* (NY), *Samuels 2427B* (NY).

Note. With its brown and coarsely striate ascospores, *N. brunneostriata* is a distinctive species but I cannot see any close relatives for it.

14. Nectria tonduzi (Spegazzini) Samuels, comb. nov. (Fig. 5A, B).

≡*Bionectria tonduzi* Spegazzini, Bol. Acad. Nac. Cienc. Córdoba **23:** 563. 1919.

Anamorph. None known.

Mycelium tan, spreading. Perithecia seated in the mycelium, solitary to cespitose in groups of ≤15, partially immersed in an erumpent, granular stroma; globose, 250–280 μm diam., nonpapillate, wall warted, pale orange, warts lighter, not changing color in 3% KOH, collapsing by lateral pinching.

Cells at surface of perithecial wall nearly circular in outline, 12–25 μm diam., walls 2–4 μm thick. Perithecial wall ca. 50 μm wide; longitudinal section comprising two regions; outer region ca. 25 μm wide, cells elliptic to circular in outline, 12–25 μm diam.; built up as warts up to 50 μm high; inner region ca. 25 μm wide, comprising flattened to fusiform, 10–15 × 2–3.5 μm cells. Perithecial apex formed of narrow, ca. 1 μm wide hyphal elements that arise from the inner perithecial wall; ostiolar canal periphysate.

Asci clavate, (62–)64–81(–85) × (11–)12.5–15.2 μm, apex broad, simple, 8-spored, ascospores 2-seriate throughout or 2-seriate above and 1-seriate below, completely filling each ascus. Ascospores fusiform, (16–)19.7–24.5(–27) × (5–)5.5–6.5(–7) μm, equally 2-celled with 1–2 guttules in each cell; not constricted at the septum, wall roughened with scattered, 0.5–1 μm diam. warts, colorless.

Habitat. On leaves of *Buettneria carthagensis* Jacq., possibly parasitizing perithecia of *Puiggarina costaricensis* Sydow.

Known Distribution. Costa Rica, known only from type.

Holotype. Costa Rica. San José. On *Buettneria carthagensis, A. Tonduz* (*1644*), x.1894 (LPS!).

Notes. *Bionectria* Speg. (Spegazzini, 1919) was originally described as a repository for species of *Nectria*-like fungi that occur on organs of living

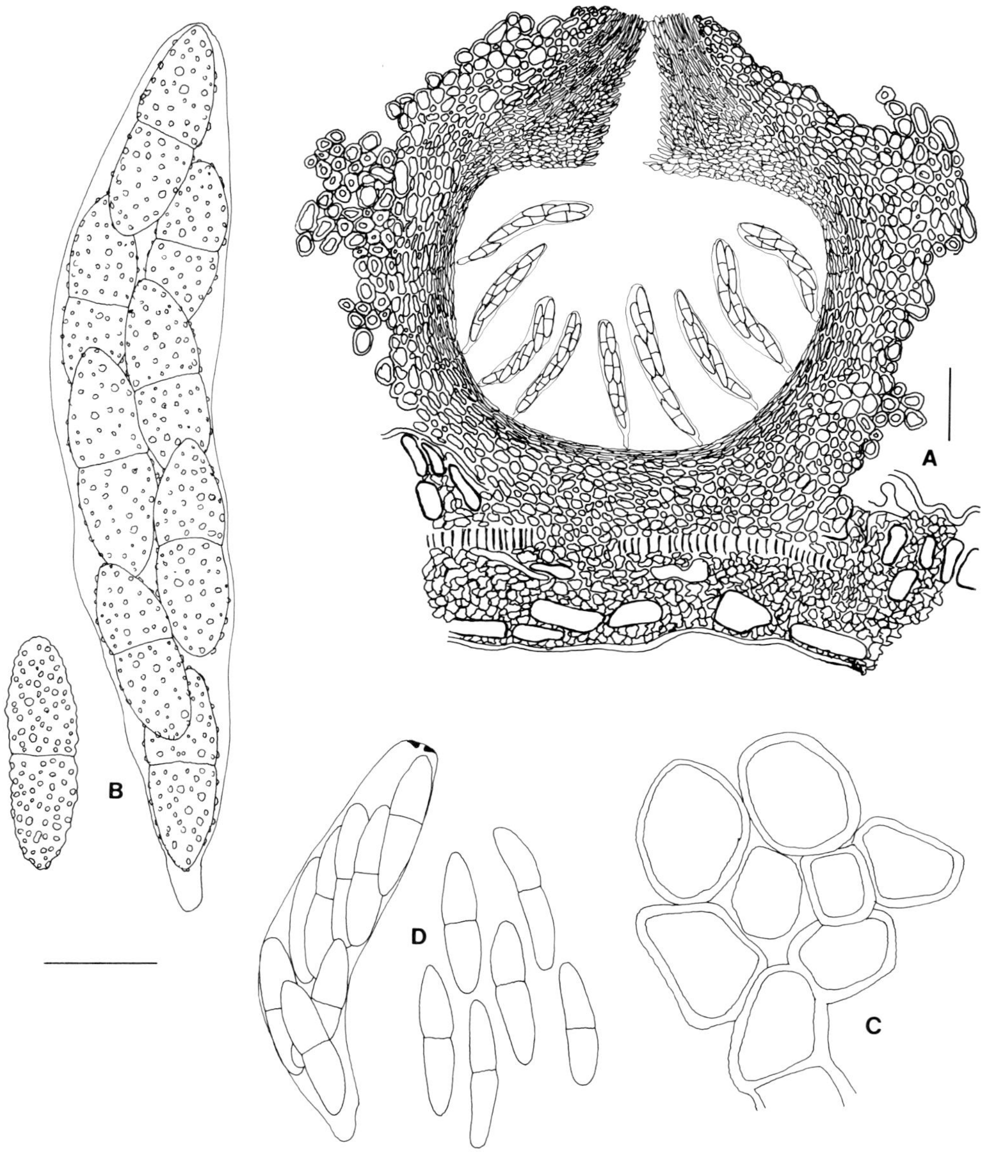

FIG. 5. A, B. *Nectria tonduzi.* **A.** Section through a mature peithecium. **B.** Ascus and ascospores. C, D. *Nectria discoidea.* **C.** Cells at surface of perithecial wall. **D.** Ascus and ascospores. All drawn from the holotypes. Lines = 10 μm.

plants. The genus has remained monotypic for *B. tonduzi.* Parasitism *per se* cannot be used as a generic character in the Hypocreales, an order where all trophic modes prevail. The orange perithecia, distinctly cellular perithecial wall and warted ascospores suggest a relationship of *N. tonduzi* to *N. ochroleuca.*

A closely related but apparently distinct unnamed species has been found on a grass leaf, possibly parasitizing a smut [Colombia: vic. km post 79 from Pasto on the Pasto–Mocoa Rd., Intendencia Putamayo, ca. 9100 ft, *Dumont (CO-3642) et al.,* 26 Jan 1976 (NY)]. Ascospores in this collection are somewhat shorter and narrower (17–21 × 4 μm) than those of *N. tonduzi*. The perithecia are associated with white sporodochia of a *Dendrodochium* sp.

15. Nectria peziza (Tode: Fries) Fries, Summa Veg. Scand. 388. 1849.

≡*Sphaeria peziza* Tode, Fungi Meckl. Sel. Fasc. **2:** 46. 1791.
≡*Sphaeria peziza* Tode: Fries, Syst. Mycol. **2:** 452. 1823.
=*Nectria consanguinea* Rehm, Hedwigia **26:** 92. 1887.
=*Nectria henningsii* Rehm, Hedwigia **28:** 352. 1889.
=*Nectria fallax* Rick, Ann. Mycol. **4:** 309. 1906.

Anamorph. *Acremonium* sp.

Mycelium not evident, or scant. Perithecia superficial, nonstromatic and easily removed from the substrate, globose, 370–420 μm high × (250–)370–430(–550) μm wide, smooth, yellow to orange, becoming cupulate when dry.

Asci clavate, (49–)60–75(–100) × (5–)8–10 (–14) μm, apex simple. Ascospores ellipsoidal, (9–)11–14(–17) × (3–)5–6 μm, equally 2-celled; striate, often appearing smooth; colorless.

Habitat. On diverse substrates, including *Polyporus* spp.

Note. For illustrations, additional synonyms, and complete descriptions see Booth (1959) and Samuels (1976a, 1976b).

16. Nectria suffulta Berk. & Curt., Jour. Linn. Soc. Bot. **10:** 378. 1868.

=*Nectria calamicola* Hennings & Nyman *in* Warburg, Monsunia **1:** 161. 1899.
=*Nectria simillima* Rehm, Hedwigia **39:** 222. 10 Aug 1900.
=*Nectria confluens* Petch, Ann. Roy. Bot. Garden Peradeniya **7(2):** 114. 1920.

Anamorph. *Acremonium* sp.

Mycelium not evident, or scant. Perithecia superficial, nonstromatic and easily removed from the substrate, globose, 310–370 μm high × (330–)400–500 μm wide, with scattered, triangular fascicles of hyphal hairs, yellow to orange and infrequently becoming brown, becoming cupulate when dry. Perithecial wall (30–)55–90 (–150) μm wide. Asci clavate, (45–)55–75(–130) μm, apex simple. Ascospores ellipsoidal, (10–)12–17(–22) × 4–5 μm, equally 2-celled, striate or appearing smooth, colorless.

Habitat. On wood and herbaceous debris, infrequently on Xylariaceae.

Note. For illustrations, additional synonymy, and a complete description see Samuels (1976a).

17. Nectria discoidea Petch, Ann. Roy. Bot. Garden, Peradeniya **7:** 126. 1920.
(Fig. 5C, D).

Anamorph. None known.

Mycelium white, thin and arachnoid, barely visible. Perithecia solitary or gregarious in groups of 3–10 on stroma of host, superficial, nonstromatic, subglobose, 120–145 μm diam., nonpapillate, becoming cupulate when dry, smooth; brown-orange, becoming orange in 3% KOH.

Cells at surface of perithecial wall circular to angular in outline, ca. 10 μm diam., walls ca. 1 μm thick. Perithecial wall ca. 45 μm wide laterally, comprising two regions. Outer region ca. 25 μm wide, comprising 2–3 layers of circular to elliptic to angular cells 5–10 μm across with walls pigmented, 1–1.5 μm thick. Inner region ca. 15 μm wide, comprising elliptic to flattened cells, ca. 10 × 3–4 μm with walls ca. 1 μm thick and nonpigmented. Ostiolar region of small cells, otherwise not anatomically distinct from the lateral wall; ostiolar canal periphysate.

Asci clavate, 30–37 × 6.5–10 μm, apex with a refractive ring; 8-spored, ascospores biseriate, completely filling each ascus. Ascospores ellipsoidal to elliptic-fusiform, (10–)12.2–14.8(–15) × (2.5–)2.8–3.2(–3.5) μm, with narrowed ends, equally 2-celled or basal cell somewhat longer than the apical cell, colorless, smooth.

Habitat. On *Xylaria* sp.

Known Distribution. Sri Lanka, known only from the type collection.

Holotype. Ceylon (Sri Lanka). Central. Hakgala, on *Xylaria* sp., *Petch 4577,* May 1913 (K!).

Note. Because of its wide perithecial wall that is composed of cells that are circular in section, and cupulate collapse, *Nectria discoidea* appears to be related to *Nectria peziza* and *N. suffulta.*

V. Nectriopsis Maire, Ann. Mycol. **9:** 323. 10 Aug 1911, *emend.* Samuels. Lectotype: *N. violacea* (Fries) Maire (*Sphaeria violacea* Fries) *vide* Weese, Z. Gärungsphysiol. **21:** 239. 1913.

Perithecia superficial or immersed in hyphae, not conspicuously stromatic, <200 μm diam., pale yellow to nearly white, KOH− (rarely violaceous and then not red in 3% KOH); <20 μm wide and comprising a single region of small thin-walled (<0.5 μm in greatest dimension), nondescript cells; cells at the surface of the wall forming *textura epidermoidea*. Fungicolous, occurring on perithecia or ascomycetous mycelium, lichenicolous, or myxomyceticolous, less frequently herbicolous.

Three intergrading subgroups may be recognized:

Va. Perithecial hairs of enlarged cells, arising from the perithecial surface.

Included species: *Nectriopsis mindoensis.*

Vb. Perithecial apex formed of a palisade of hyphal elements that intergrade with the periphyses; short outgrowths or long and conspicuous hyphal hairs sometimes arising from elements of the papilla.

Included species: *Nectriopsis candicans, N. cupulata, N. discicola, N. discophila, N. exigua, N. guarapiensis, N. hirsuta, N. hyperbiota, N. hypocrellicola, N. lasioderma, N. lasiodermopsis, N. leucorrhodina, N. microthecia, N. oidioides, N. oropensoides, N. ostiolorum, N. perpusilla, N. pipericola, N. porinicola, N. septofusidiae, N. sepultariae, N. sibicola, N. sororicola, N. sporangiicola, N. violacea.*

Vc. Perithecial apex formed of small cells and not anatomically distinct from the perithecial wall; outgrowths from around the ostiolar opening not formed.

Included species: *Nectriopsis albofulta, N. bactridioides, N. byssotecta, N. cephalosporii, N. cordiae, N. epichloe, N. epinectria, N. flavella, N. lilliputia, N. macroepichloe, N. nanocarpa, N. oxyspora, N. puggarii, N. squamulosa.*

The three subgroups intergrade into each other and the anatomy of the perithecial apex can be difficult to see unless microtome sections are studied. I do not feel that these subgroups represent fundamentally different organisms but rather that they offer a more or less convenient way of breaking up a large group of species.

Nectriopsis is easily distinguished from *Nectria sensu stricto.* The type species of *Nectria, N. cinnabarina* and its closest relatives have large, stromatic, red perithecia and distinctive sporodochial, synnematous or pycnidial anamorphs (*Tubercularia* spp., see Seifert, 1985). These species are parasitic or saprobic on many woody, dicotyledonous plants. It is more difficult to distinguish *Nectriopsis* from species of *Nectria* that have yellow to orange perithecia (e.g., *N. ochroleuca*). The thin perithecial wall with *textura epidermoidea* at the surface, combined with yellow pigmentation, small perithecia and a fungicolous habit are the principal characters that distinguish *Nectriopsis* from groups of *Nectria* species that have larger perithecia with thicker perithecial walls and a habit that is not conspicuously fungicolous. The lectotype species of *Nectriopsis, N. violacea* has violaceous perithecia and occurs on the slime mold *Fuligo septica* (L.) Weber. The violaceous pigmentation of perithecia of the type species is unfortunate because it isolates the species within the genus, but the perithecial anatomy and habitat confirm its affinities to the remaining species that have white to yellow perithecia. No pigment is released from perithecia of *N. violacea* when immersed in potassium hydroxide; perithecia of *Gibberella* species are purple, and a red pigment is often released in potassium hydroxide. I have included eight species (*Nectriopsis mindoensis, N. sororicola, N. hypocrellicola, N. epimyces, N. puiggarii, N. flavella, N. cordiae, N. macroepichloe*) that have a perithecial wall greater than 20 μm wide. The wall of these species (except *N. flavella*) comprises a single region of cells as do the walls of more typical species. The overall aspect of these species conforms well with my concept of *Nectriopsis.* The perithecial wall of *Nectriopsis flavella* comprises two regions of thin-walled cells; the species may be more closely allied to *Nectria ochroleuca* but in the absence of information about the anamorph of *N. flavella,* I prefer to retain the species in *Nectriopsis* because of its overall similarity to other species included in the genus.

Some species of *Nectriopsis* are not conspicuously fungicolous, but the perithecia of these species form on herbaceous debris that is colonized by other fungi and so may be growing on mycelium.

Few species of *Nectriopsis* have been positively linked to anamorphs and the majority of those anamorphs are nondescript, neither confirming nor denying a relationship among the species. Most of the proven or associated anamorphs have mononematous, monophialidic conidiophores. Conidia of most species are colorless (green in *N. epichloe*), unicellular, ellipsoidal and less than 10 μm long but conidia of two species are phragmosporous (*N. sororicola, N. sibicola*). Conidia of most species are held in drops of liquid, but conidia of *N. septofusidiae* and *N. epimyces* are held in dry chains. Often teleomorphs of two or more collections indicate a single species while

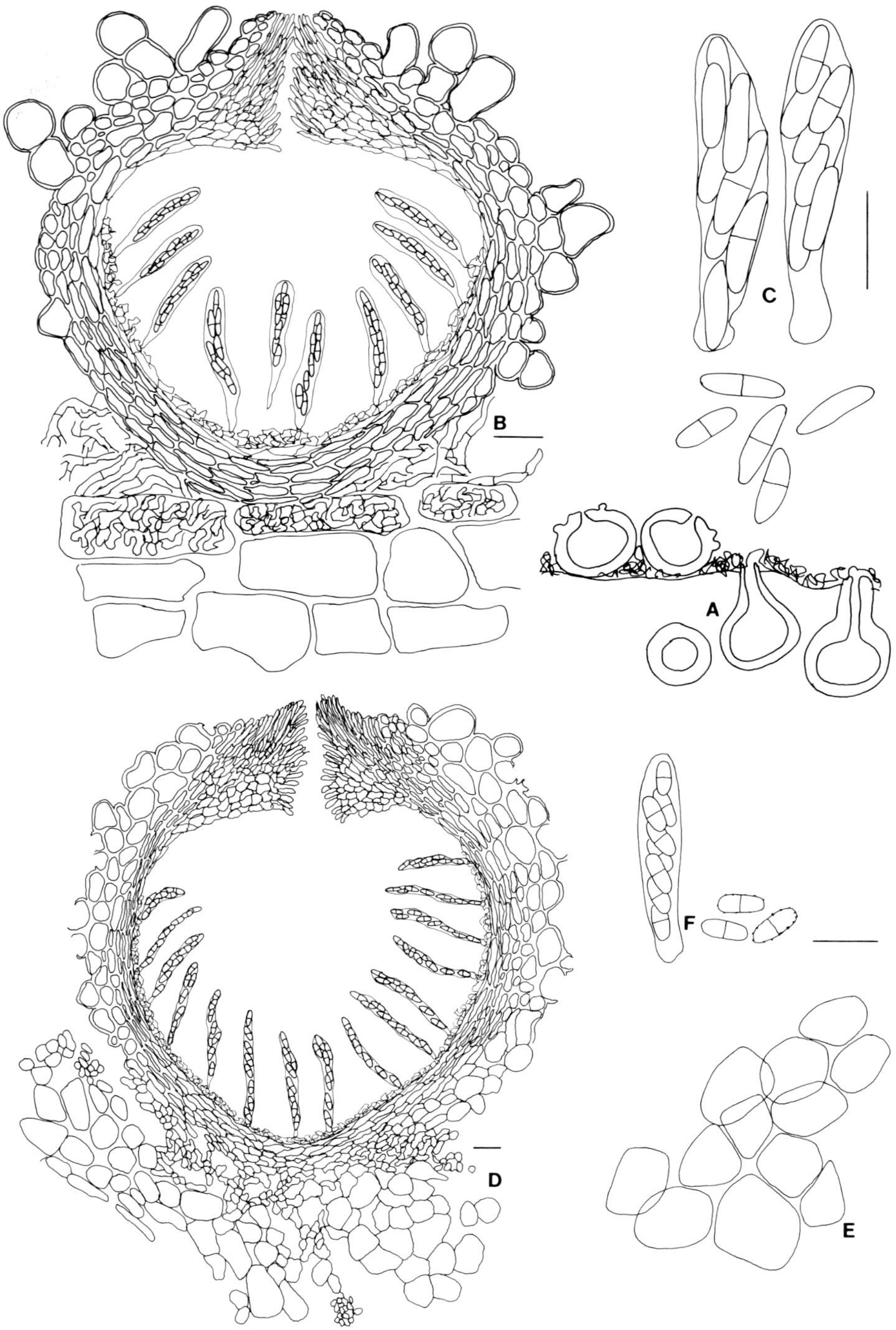

FIG. 6. A–F. *Nectriopsis mindoensis.* **A.** Habit sketch of two superficial perithecia and three immersed, parasitized perithecia. **B.** Section through a mature perithecium. **C.** Asci and ascospores. **D.** Section through a mature perithecium of New York collection. **E.** Cells at the surface of the perithecial wall of New York collection. **F.** Ascus and ascospores of New York collection. A–C drawn from the holotype; D–F drawn from *Rogerson 2476.* Lines = 10 μm.

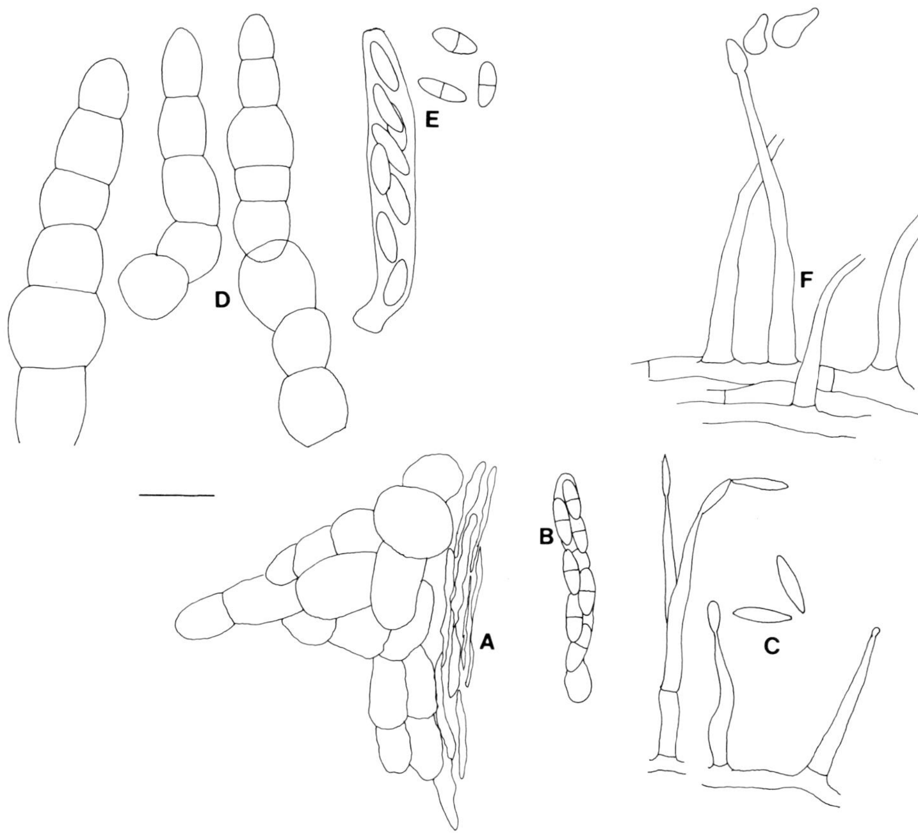

FIG. 7. A–F. *Nectriopsis mindoensis,* Colombian collections. **A.** Portion of the wall of a mature perithecium showing cells built up to form a conical hair. **B.** Ascus with ascospores. **C.** Associated conidiophores and conidia. **D.** Perithecial hairs. **E.** Ascus and ascospores. **F.** Associated conidiophores and conidia. A–C drawn from *Dumont-CO 7410*; D–F drawn from *Dumont-CO 6578.* Lines = 10 μm.

the anamorphs found on the host in association with these teleomorphs indicate more than one species (e.g., *N. mindoensis, N. epimycota, N. squamulosa*).

Maire (1911) originally considered that *Nectriopsis* approached *Hypomyces* (Fr.) Tul. because of the fungicolous habit and the formation of a subiculum. I do not consider *Nectriopsis* to be closely related to *Hypomyces.* Perithecia of several species of *Nectriopsis* form in mycelium but this mycelium is not ontogenetically comparable to the subiculum of *Hypomyces* species.

Nectriopsis and *Hypomyces* are similar in their fungicolous habit but the Hypocreales is rife with fungus-inhabiting species. The present work deals with a limited number of species. No doubt many others remain to be discovered. Species that parasitize meliolaceous leaf parasites (Samuels, 1976a), discomycetes (Rogerson & Samuels, 1985) and myxomycetes (Samuels, 1973) are briefly described here but illustrations and complete descriptions are provided in the cited references.

Gams and Van Zaayen (1982) used the genus *Nectriopsis* for various species that occur on basidiocarps and that have *Verticillium* anamorphs. With the exception of *Nectriopsis exigua,* a parasite of slime molds, I do not think these species are congeneric with *Nectriopsis violacea* and suggest that their affinities lie with *Sphaerostilbella* Hennings (see Seifert, 1985).

18. Nectriopsis mindoensis (Petrak) Samuels, comb. nov. (Figs. 6, 7).

≡*Nectria mindoensis* Petrak, Sydowia **4:** 509. 30 Sep 1950.

Anamorph. ?*Acremonium* sp.

Mycelium thin and white and spreading, or not evident. Perithecia densely gregarious, superficial, nonstromatic, sometimes seated on a thin layer of densely compacted hyphae and difficult to remove from the substrate, globose, 95–200(–260) μm diam., nonpapillate, not collapsing when dry, or collapsing by lateral pinching, less frequently becoming cupulate when dry, yellow, not changing color in 3% KOH; wall slightly roughened, often with hairs consisting of globose to ellipsoidal cells 7–15 μm with walls 1–2 μm thick, solitary and often stalked or joined in chains of up to three arising from the upper half of the perithecial wall.

Cells at surface of perithecial wall pseudoparenchymatous, 5–10 μm in greatest dimension, walls 0.5–1 μm thick. Perithecial wall ca. 15 μm wide laterally and comprising a single region of elliptical to ± circular cells 5–8 × 3–4 μm, walls 0.5–1 μm thick; cells becoming progressively more flattened and compressed toward the perithecial locule; sometimes with a 20–35 μm wide region around the perithecial apex, cells of the mantle ± circular in outline, 7–10 μm diam., walls ≤0.5 μm thick; ostiolar mantle formed of narrow hyphal elements ca. 1 μm wide, arising from inner cells of the perithecial wall and merging with the periphyses.

Asci clavate, (29–)31–47(–60) × 5.5–8(–9) μm, apex simple; 8-spored, ascospores biseriate, completely filling each ascus. Ascospores ellipsoidal to oblong, (5–)6.7–10.6(–12.5) × (1.5–) 2.2–3 μm, equally 2-celled, not constricted, smooth to finely spinulose, colorless.

Habitat. On *Nectria peziza,* immersed ascomycetes, or lichens (*Dimerella* sp.), or not obviously fungicolous or lichenicolous.

Known Distribution. Colombia, Ecuador, Norway, Panama, USA (New York).

Holotype. Ecuador. Prov. Pichincha. Mindo, on Bambusae, *H. Sydow 391,* 27 Oct 1937 (W!).

Additional Specimens Examined. PANAMA. VERAGUAS: Lower slopes of Cerro Tuté, ca. 8 km NE of Santa Fé on property of Agricultural School Alto de Piedra, 2000 ′, on rachis of indet. fern, *Dumont (PA 648), Carpenter & Mori,* 19 Jun 1975 (NY). COLOMBIA. ANTIOQUIA: Ca. 1 km off the Medellín–Urama Rd. at a point 83 km from Medellín, oak forest Chaquire, ca. 9200 ft, on *Dimerella* sp., *Dumont (CO 6578), Sherwood & Velásquez,* 17 Aug 1976 (NY). CHOCÓ: Ca. 66 km from Ansermanuevo, on the Ansermanuevo–San José del Palmar Rd., 1900 m, on indet. herbaceous stem, *Dumont (CO 7410), Molina & Forero,* 27 Aug 1976 (NY). Cundinamarca: Ca. 15 km from Bogotá on the Bogotá–Choachi Rd., ca. 10,100 ft, on *Puya* leaves, *Dumont (CO 5535), Carpenter & Sherwood,* 4 Aug 1976 (NY). NORWAY. FINNMARK: Vardö, beneath the old fortress Vardö Hus, on *Nectria* sp. on overwintered stems and petioles of *Geum rivale* L., *K. & L. Holm 4262b,* 27 Aug 1986 (NY, UPS). USA. NEW YORK: Schuyler County, Arnot Forest of Cornell University, on cut end of stump and on *Nectria peziza, Rogerson 2476,* 2 Aug 1948 (NY).

Notes. No perithecia were found on an isotype of this species preserved at ZT.

Nectriopsis mindoensis is most easily recognized by the globose, often catenate, cells of the perithecial apex. This unusual feature is common to the South American and Norwegian collections cited above. Perithecia of the New York collection lack these hairs; instead, there is a mantle of more or less globose cells around the perithecial apex. In other respects the New York collection is typical of the species.

Two additional collections, from Colombia, also differ from typical *N. mindoensis* in subtle details. In both of these collections the circular cells of the wall tend to form hair-like chains 20–50 μm long that stand erect from the surface of the perithecial wall. Perithecia of *Dumont-CO 7410* (Fig. 7A–C) have developed on perithecia of *Nectria suffulta*; ascospores measure (5.5–) 6–7.5 × 1.7–2.2 μm and are smooth and colorless. An associated *Acremonium*-like anamorph has ellipsoidal conidia, 7.5–9 × 1.8 μm, that are held end to end in chains. Perithecia of the collection *Dumont-CO 6578* (Fig. 7D–F) are seated on fructifications and thallus of a lichen (*Dimerella* sp.); ascospores are (5.5–)6.2–8(–10) × (1.8–)2–2.8(–3) μm and are smooth and colorless. The associated *Acremonium* anamorph has pyriform conidia 6.5–7.5(–8) × (1.8–)2–2.8(–3) μm. The perithecia of these two collections are certainly anatomically very close to each other; they differ primarily in the habitats, less well developed hairs and smaller ascospores in *CO 7410.* The associated anamorphs strongly suggest that each collection is of a distinct species.

19. Nectriopsis perpusilla (Montagne) Samuels, comb. nov. (Fig. 8A, B).

≡*Hypocrea perpusilla* Montagne *in* Sagra, Hist. fis. pol. nat. Cuba ix, 335. 1838–1842 [1842].

≡*Nectria perpusilla* (Montagne) Montagne, Syll. gen. spec. Crypt. 225. 1856 *non* Saccardo 1918.

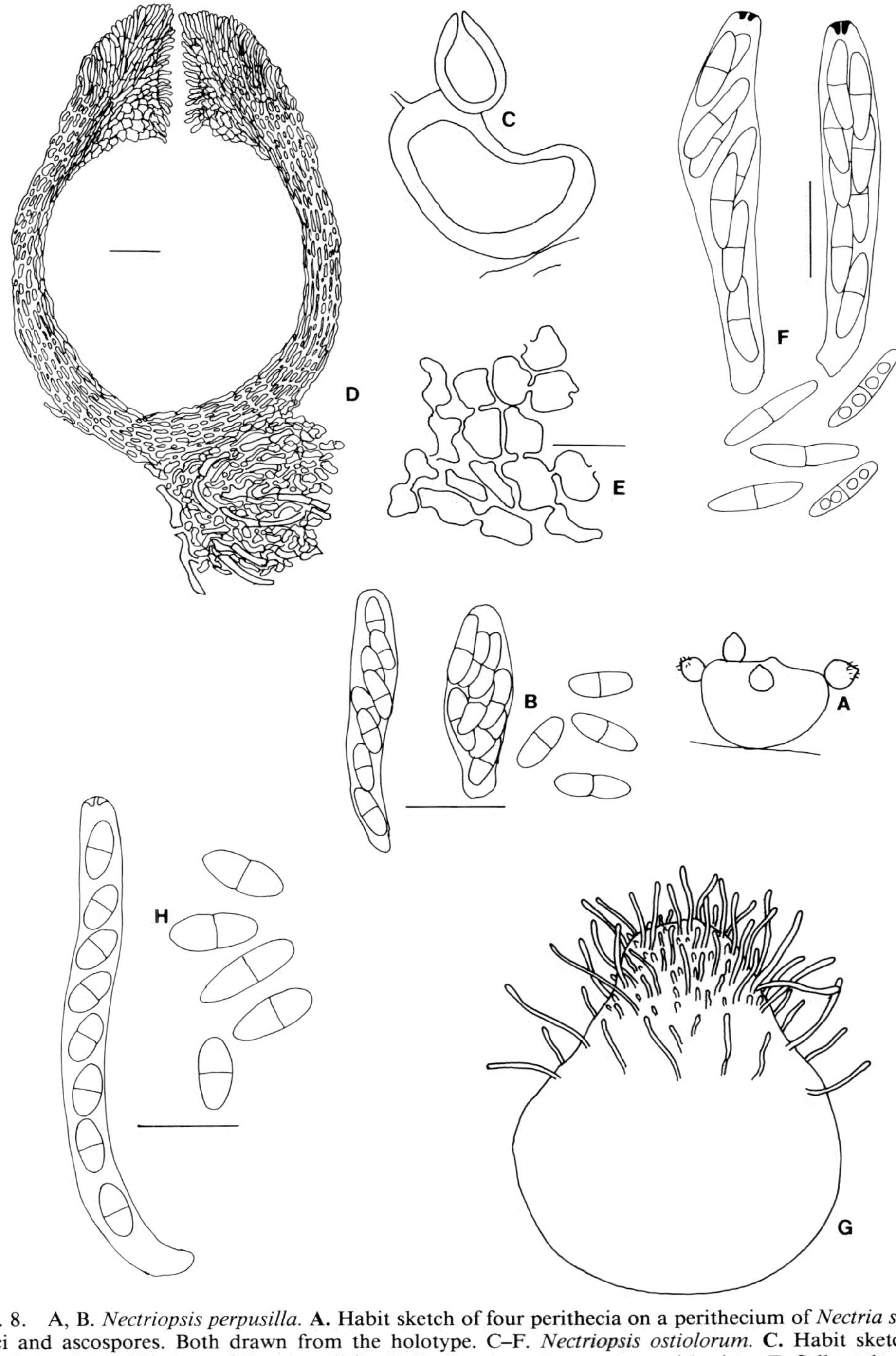

FIG. 8. A, B. *Nectriopsis perpusilla.* **A.** Habit sketch of four perithecia on a perithecium of *Nectria suffulta.* **B.** Asci and ascospores. Both drawn from the holotype. C–F. *Nectriopsis ostiolorum.* **C.** Habit sketch of a perithecium on a perithecium of *Nectria suffulta.* **D.** Section through a mature perithecium. **E.** Cells at the surface of the perithecial wall. **F.** Asci and ascospores. A–F drawn from *Dumont-EC 1673A.* G, H. *Nectriopsis lasioderma.* **G.** Sketch of a mature perithecium with hairs. **H.** Ascus and ascospores. G, H drawn from the holotype. Lines = 10 μm.

≡*Nectriella perpusilla* (Montagne) Saccardo, Michelia **1:** 51. 1 Jun 1877.
=*Lisea parasitica* Rick, Broteria **5:** 41. 1906.

Anamorph. None known.

Mycelium not apparent. Perithecia solitary or cespitose in groups of up to 10, superficial, nonstromatic, pyriform, (85–)106–163 μm diam., with a blunt papilla, smooth, collapsing by lateral pinching when dry, pale yellow to nearly colorless, not changing color in 3% KOH; with white, unbranched, septate, hyphal, hairs 10–20 μm long × 3–4 μm wide at base, wall 1 μm thick, forming a more or less obvious fringe around the perithecial apex depending on the collection.

Cells at the surface of the perithecial wall forming *textura epidermoidea,* 5–7 μm wide, walls unevenly thickened, lumina of some adjacent cells joined by fine pores. Perithecial wall ca. 10 μm wide, comprising a single region of elliptical, 5–7 × 2–3 μm cells with thin walls; perithecial apex formed by a compact palisade of hyphal elements, each with a rounded tip and many growing outwardly to form the hairs, becoming progressively narrower and more thin-walled toward the ostiolar canal and merging there with the periphyses.

Asci clavate, 25–34(–40) × 4–6(–7) μm, apex simple (type collection) or with a minute ring; 8-spored, ascospores biseriate, completely filling each ascus. Ascospores ellipsoidal to oblong, (5–)5.7–7.3(–9) × 2–2.5(–3) μm, equally 2-celled, constricted or not at the septum, smooth or finely spinulose, colorless (type collection) or pale brown.

Habitat. On perithecia of *Hypoxylon, Nectria,* and *Xylaria.*

Known Distribution. Brazil, Colombia, Cuba, New Zealand, Venezuela.

Holotype. Cuba: Ad *Sphaeriam allantodium* Berk. (Hist. Nat. Cuba, herb. Montagne: K. presumed holotype!).

Additional Specimens Examined. COLOMBIA. ANTIOQUIA: Ca. 37 km from Dabeiba, on the Dabeiba–Turbo Rd., ca. 800 ft, on *Nectria suffulta, Dumont (CO 6502), Carpenter & Sherwood,* 15 Aug 1976 (NY). CUNDINAMARCA: Vic. km 15 from Fomeque, rd. between Calera and Fomeque, on *Nectria* cf. *cyatheae* Dingley on internodes of *Chusquea* sp., *Dumont (CO 978) et al.,* 10 Jul 1974 (NY). VENEZUELA. DTO. FEDERALE: Mt. La Naiguata, above La Planta, trail between Río Chiquito and Río Grande, on *Nectria suffulta* on log, *Dumont (VE 747), Haines & Manara,* 25 Jun 1971 (NY, VEN). BRAZIL. RIO GRANDE DO SUL: São Leopoldo, on *Hypoxylon, Rick,* 1904 (PACA 13997 [1502], HOLOTYPE of *Lisea parasitica*). NEW ZEALAND. NORTHLAND: Hokianga County, Mangamuka Bridge, vic. Omahuta Forest Headquarters, on *Nectria* sp., *Samuels,* 11 May 1983 (PDD 46285).

Notes. The various specimens assigned herein to *N. perpusilla* differ subtly from the type collection of the species. The basic concept of *N. perpusilla* is of a species with small yellow perithecia seated directly on the perithecium of the host ascomycete and with no obvious mycelium; with short and often inconspicuous hyphal hairs that grow outwardly from the perithecial papilla (although none were noted on perithecia of the type collections of *N. perpusilla* or *Lisea parasitica*); ascospores that are among the smallest found in the entire genus [(6–)6.5–7.5(–8) × 2–2.8(–3) μm in the type collection of *N. perpusilla*] and that are smooth and colorless. Ascospores of *CO 978* are pale brown, and ascospores of *VE 747* are finely spinulose.

Although the type specimens of fungi described by Montagne are usually found in PC, I was not able to locate any material of *Hypocrea perpusilla* there. The label on the specimen found in K indicates that this collection is at least a part of the original gathering.

20. Nectriopsis sepultariae (Ade) Samuels, comb. nov.

≡*Hypomyces sepultariae* Ade, Hedwigia **64:** 304. 29 Dec 1923.
≡*Nectria sepultariae* (Ade) Petrak, Ann. Mycol. **27:** 335. 10 Dec 1929.

Anamorph. None known.

Mycelium not evident. Perithecia pyriform, 275–310 × 210–245 μm, with a blunt apex; red-orange, not changing color in 3% KOH, cells at surface of perithecial wall angular. Asci broadly cylindrical, (85–)95–140(–145) × 18–23(–25) μm, apex simple. Ascospores broadly ellipsoidal to fusiform, (23.5–)24.3–27.7(–29) × (10–)10.6–12.7(–13) μm, colorless, smooth.

Habitat. Forming directly on the hymenium of the operculate discomycete *Sepultaria arenicola* (Lév.) Rehm.

Notes. *Nectriopsis sepultariae* is unusual in the genus because of the red-orange perithecial color. The red coloration does not intensify in 3% KOH, therefore I believe it is merely a variation of the usual yellow perithecial coloration in *Nectriopsis.*

For illustrations and a complete description see Rogerson and Samuels (1985).

21. Nectriopsis discophila (Rogerson & Samuels) Samuels, comb. nov.

≡*Nectria discophila* Rogerson & Samuels, Mycologia **77:** 777. 15 Oct 1985.

Anamorph. ?*Acremonium* sp.

Mycelium white, cottony. Perithecia globose to broadly pyriform, (170–)200–250(–300) × (150–)170–200(–260) μm, amber colored when dry, pale to bright yellow with a faintly reddish apex in water and 3% KOH. Asci clavate, 32–50(–55) × (5–)6–10(–12) μm, apex simple. Ascospores fusiform with acute ends, 11–20 × 2–3 μm, equally 2-celled, verruculose to smooth, colorless.

Habitat. Forming in white mycelium on apothecia of *Lachnum* sp.

Note. For illustrations and a complete description see Rogerson and Samuels (1985).

22. Nectriopsis discicola (Rogerson & Samuels) Samuels, comb. nov.

≡*Nectria discicola* Rogerson & Samuels, Mycologia **77:** 782. 15 Oct 1985.

Anamorph. None known.

Mycelium white, hyphae erect, straight, smooth, unbranched. Perithecia pyriform, 140–200 × 145–220 μm, with a blunt to acute apex with a fringe of hairs around the perithecial apex, pale yellow, not changing color in 3% KOH. Ascospores broadly ellipsoidal, (11–)11.6–14.2(–16) × (6–)6.7–7.7(–8) μm, equally 2-celled, verruculose, pale brown.

Habitat. Forming on the hymenium of *Bioscypha* sp. on the fern *Lophosoria quadripinnata* (Gmel.) C. Chr.

Note. For illustrations and a complete description see Rogerson and Samuels (1985).

23. Nectriopsis ostiolorum (Berkeley & Cooke) Samuels, comb. nov. (Fig. 8C–F).

≡*Dialonectria ostiolorum* Berk. & Cooke, Grevillea **13:** 8. Sep 1884.
≡*Nectria ostiolorum* (Berk. & Cooke) P. A. Saccardo & P. Sydow, Syll. Fung. **4:** 964. 15 Sep 1891.

Anamorph. None known.

Mycelium not apparent or thin, white, restricted to the host perithecia. Perithecia gregarious, superficial, nonstromatic but difficult to remove from the substrate, pyriform, ca. 400 μm high × ca. 225 μm wide, with a short, acute papilla, collapsing by lateral pinching when dry, smooth, pale orange, not changing color in 3% KOH.

Cells at the surface of the perithecial wall forming *textura angularis* to *textura epidermoidea* with unevenly thickened walls and pores joining the lumina of adjacent cells. Perithecial wall ca. 15 μm wide, comprising a single region of flattened, compressed cells 10–15 μm long × 2–3 μm wide, walls slightly thickened. Ostiolar region comprising hyphal elements that form a palisade and merge with periphyses.

Asci narrowly cylindrical, ca. 85 × ca. 4 μm, apex with a refractive ring; 8-spored, ascospores 1–2-seriate, completely filling each ascus. Ascospores fusiform to ellipsoidal, 13–15 × 2–3 μm, equally 2-celled or with the upper cell somewhat broader and shorter than the lower cell, not constricted at the septum, smooth, colorless.

Habitat. On perithecia of *Xylaria* sp., *Nectria* sp. and ?*Chaetosphaeria* sp.

Known Distribution. Colombia, Cuba, Ecuador.

Holotype. Cuba: On *Xylaria rhopaloides* Mont., *Wright 592,* date unknown (K!).

Additional Specimens Examined. COLOMBIA. NARIÑO: Pasto, Páramo Bordoncillo, km 33–34 E of Pasto, on the rd. to Mocoa, 3100 m, on pyrenomycete on bamboo, *Dumont (CO 9454 A), Oberwinkler & Pulido,* 6 Jul 1978 (NY). ECUADOR. EL ORO: Ca. 14 km from Pinas, toward Machalu, on the Arenillas–Loja Rd., 3700 ft, on perithecia of *Nectria (Dumont-EC 1673), Dumont (EC 1673A), Carpenter & Buriticá,* 28 Jul 1975 (NY).

24. Nectriopsis lasioderma (Ellis) Samuels, comb. nov. (Fig. 8G, H).

≡*Nectria lasioderma* Ellis, Am. Nat. **17:** 194. Feb 1883.
=*Nectria infusaria* Cook & Harkness *in* Cooke, Grevillea **12:** 101. Jun 1884.
≡*Nectriopsis infusaria* (Cooke & Harkness) Doi, Mem. Nat. Sci. Mus. Tokyo **11:** 17. 20 Dec 1978.

Anamorph. None known.

Mycelium not evident. Perithecia solitary to gregarious, superficial or with base immersed in host perithecium, nonstromatic, pyriform 175–190 μm high × 140–190 μm wide, with an obtuse to acute papilla, collapsing by lateral pinching, smooth, pale yellow to nearly white, not changing color in 3% KOH; with a fringe of hairs arising from the upper third of the perithecial wall, growing outwardly from cells of the perithecial papilla; hairs white, hyphal, unbranched, straight, septate, 20–60 μm long × 2–3 μm wide, wall ≤0.5 μm thick.

Perithecial wall covered with ca. 2 μm wide hyphae with wall ≤0.5 μm thick. Perithecial wall ca. 15 μm wide, comprising a single region of flattened cells ≤10 μm long × 2–3 μm wide; perithecial apex formed of a palisade of 2–3 μm wide hyphal elements with rounded ends and that grow outwardly to form the perithecial hairs; merging with the periphyses within.

Asci cylindrical, (62–)66–84(–95) × (5–)7–10 μm, apex with a refractive ring; 8-spored, ascospores 1-seriate with overlapping ends, completely filling each ascus. Ascospores ellipsoidal, (8–)9.5–13.5(–17) × (3.5)4.2–6(–7) μm, equally 2-celled, not constricted at the septum, smooth, pale brown or colorless.

Habitat. On perithecia and anamorphs of various pyrenomycetes including *Apiosporina, Nectria,* and *Valsa.*

Known Distribution. Brazil, USA (California, Connecticut, New Jersey).

Holotype. USA. New Jersey: Newfield, on *Valsa lutescens* Ellis, collector unknown, Jun 1882 (NY!).

Additional Specimens Examined. USA. CALIFORNIA: On *Acacia* sp. with *Fusarium acaciae* Cooke & Harkn., *Harkness 2268,* date not known (lectotype of *Nectria infusaria vide* Doi: BPI, K); second colleciton, same data except *Harkness 2362* (syntype of *Nectria infusaria*: BPI, K). CONNECTICUT: Litchfield County, SW edge of Litchfield, White Memorial Foundation, environmental center, on *Apiosporina morbosa* (Schw.) Arx on *Prunus* sp., *Rogerson 78-185,* 1 Oct 1978 (NY). NEW JERSEY: Newfield, on *Valsa lutescens* on *Quercus coccinea* Wangenh., collector unknown, Jun 1882 (North American Fungi 1177: NY). BRAZIL. SANTA CATARINA: Blumenau, on *Nectria blumenaviae* Rehm (=*N. pityrodes* (Mont.) Mont.), *Ule,* 9.1888 (S, with type collection of *N. blumenaviae*).

Notes. *Nectriopsis lasioderma* differs from *N. lasiodermopsis* in having smooth and somewhat broader ascospores, and longer perithecial hairs. An anamorph was evident on three of the four specimens of *N. lasiodermopsis* while none was found with any of the cited specimens of *N. lasioderma.*

Doi (1978) considered *Nectria infusaria* and *N. erinacea* Starb. to be synonymous. I have studied the type specimens of both species and disagree. Perithecia in the type collection of *N. infusaria* have a fringe of setae arising from around the perithecial apex, the perithecia are nonstromatic and the ascal apex has a refractive ring. Perithecia in the type collection of *N. erinacea* are glabrous and arise from a pseudoparenchymatous to prosenchymatous stroma that overgrows perithecia of *N. pityrodes*; the ascal apex is simple.

Perithecia of *N. infusaria* match those from the type collection of *N. lasioderma* except that perithecial hairs of *N. lasioderma* are longer than those of *N. infusaria.*

In his discussion of *N. infusaria,* Doi (1978) illustrated ascospores from the type collections of *N. erinacea* (S) and *N. infusaria* (K) as being markedly spinulose. Ascospores in the type collection of *N. infusaria* are considerably smaller than those given by Doi (1978) for this species. When I examined these specimens I found only smooth ascospores. Doi recollected a species in Japan that he identified as *Nectriopsis infusaria* occurring on *Nectria pityrodes.* His isolate produced a *Stachylidium*-like anamorph in pure culture. The dried culture preserved with that specimen is typical of *N. pityrodes.* It is likely that he mistakenly isolated conidia of *N. pityrodes,* thinking that they were ascospores of the hyperparasite. The perithecia of the hyperparasite that he illustrated lacked hairs and the ascospores were spinulose. This collection may represent an undescribed species. The Brazilian collection cited above has somewhat longer and narrower ascospores than were found in the type collection of *N. lasioderma.*

25. Nectriopsis lasiodermopsis Samuels, sp. nov. (Fig. 9).

Nectriopsidi perpusillae (Montagne) Samuels similis sed ascosporis majoribus, (8–)9–11 (–12.5) × (7–)9–10 μm; *N. lasiodermae* (Ellis) Samuels similis sed ascosporis angustioribus.

Status anamorphicus *Acremonium* sp.
Holotypus. New Zealand (PDD 49502).
Anamorph. *Acremonium* sp.

Mycelium not evident; conidiophores arising directly from the substrate, macronematous, mononematous, arising from 2–3 μm wide hyphae, 45–90 μm long × 2 μm wide at base, 1–3-septate, unbranched, straight, smooth, 2 μm wide at base, monophialidic; phialides 40–70 μm long, tip 1–1.5 μm wide; collarette at most slightly thickened, not flared. Conidia 0–1-septate; 0-septate: (5–)6.5–10.5(–12) × 2–3 μm; 1-septate: 11–14.5(–16) × 2.5–3 μm; cylindrical (1-

septate) to ellipsoidal (0-septate), lacking an obvious basal abscission scar; held in colorless drops of liquid.

Perithecia densely gregarious, superficial, nonstromatic, difficult to remove from the substrate, pyriform, 185–200 × 155–180 μm, with a short papilla, smooth, collapsing by lateral pinching when dry, pale yellow, nearly white, not changing color in 3% KOH; with short, 10–20 × 4–5 μm hairs protruding from the papilla.

Cells at the surface of the perithecial wall forming *textura epidermoidea* with unevenly thickened walls, lumina of adjacent cells joined by fine pores. Perithecial wall ca. 15 μm wide, comprising a single region of indistinct cells; perithecial papilla formed of a palisade of hyphal elements, many of which grow outwardly to form hairs.

Asci clavate, (35–)37–45(–48) × (7–)9–10 μm, apex with a frequently inconspicuous refractive ring; 8-spored, ascospores biseriate, partially to completely filling each ascus. Ascospores ellipsoidal, (8–)9–11(–12.5) × 3–4 μm, equally 2-celled, not constricted at the septum, finely spinulose to smooth, colorless.

Characteristics in Culture. Colonies on CMD very slow growing, white, with scant aerial mycelium, surface of colony covered with slime. Conidiophores as found in nature; conidia formed in culture larger than those found in nature [0-septate: (7–)9–15(–17) × 2–3 μm; 1-septate: (11–)13.5–20.5(–23) × (2.5–)2.8–3.7(–4) μm]; relatively few 0-septate conidia found in culture.

Habitat. On perithecia of *Nectria* spp.

Known Distribution. New Zealand, Australia (Tasmania).

Etymology of the Specific Epithet. Refers to a similarity to *Nectriopsis lasioderma*.

Holotype. New Zealand. North Island. Gisborne: Urewera National Park, Lake Waikaremoana, Ngamoko Track, on old *Nectria* cf. *ochroleuca, Samuels, Horak, Johnston & Petersen,* 21 May 1981 (PDD 49502: isotype NY).

Additional Specimens Examined. AUSTRALIA. TASMANIA: Vic. Hobart, Mt. Wellington, nr. Fern Tree, on *Nectria, Doi s.n.,* 23 May 1986 (NY); same collecting data, second collection, *Doi D-6761* (NY); Mt. Field National Park, on *Nectria* sp., *Doi D-6766,* 29 May 1986 (NY).

Notes. The specimens of *N. lasiodermopsis* from Tasmania concur well with the New Zealand collection and there is no doubt that all these collections are *N. lasiodermopsis.* The anamorph found on the type and one of the Tasmanian collections (no anamorph was seen on the other two collections) is close to the anamorph formed in cultures derived from single ascospores. Unfortunately, those cultures are no longer viable.

Nectriopsis lasiodermopsis is close to *N. perpusilla* and *N. lasioderma.* It differs from *N. perpusilla* in having larger, usually spinulose ascospores; from *N. lasioderma* in having shorter perithecial hairs and narrower, usually spinulose ascospores.

26. Nectriopsis sororicola Samuels, sp. nov. (Fig. 10).

Nectriopsidi perpusillae (Mont.) Samuels similis, sed ascosporis striatae, (8–)8.5–11(–12) × (3–)3.7–4.7(–5) μm.

Status anamorphicus *Acremonium*-similis.
Holotypus. New Zealand (PDD 41423).
Anamorph. *Acremonium*-like.

Mycelium not evident. Perithecia solitary, superficial, nonstromatic, anchored to substrate by a fringe of white hyphae, pyriform 190–280 (–375) × 150–250(–310) μm, with an acute papilla, smooth, collapsing by lateral pinching when dry; pale yellow, not changing color in 3% KOH.

Cells at the surface of the perithecial wall forming *textura angularis* to *textura epidermoidea,* with unevenly thickened walls. Perithecial wall 15–30 μm wide, comprising a single region of elliptic, 5–7 μm long, indistinct cells. Perithecial papilla formed of a palisade of hyphal elements with tip cells, at exterior, clavate to circular in outline, 7–10 × 5–7 μm, walls ca. 0.5 μm thick; hyphal elements becoming progressively more narrow and thin-walled toward the ostiolar canal and there merging with the periphyses.

Asci cylindrical to clavate, (63–)67–75(–90) × 6–8(–9) μm, apex with a ring; 8-spored, ascospores 1-seriate throughout or 2-seriate above and 1-seriate below. Ascospores ellipsoidal, (8–)8.5–11(–12) × (3–)3.8–4.7(–5) μm, equally 2-celled, not constricted at the septum, coarsely striate, colorless.

Characteristics in Culture. Colonies grown two weeks at 20°C, 12 hours near ultraviolet + cool white fluorescent light. CMD, PDA: barely growing away from the original inoculum, colony dense, velvety, pale salmon with a strong actinomycetous odor. Conidia forming on CMD and PDA; conidiophores arising from 2–3 μm wide

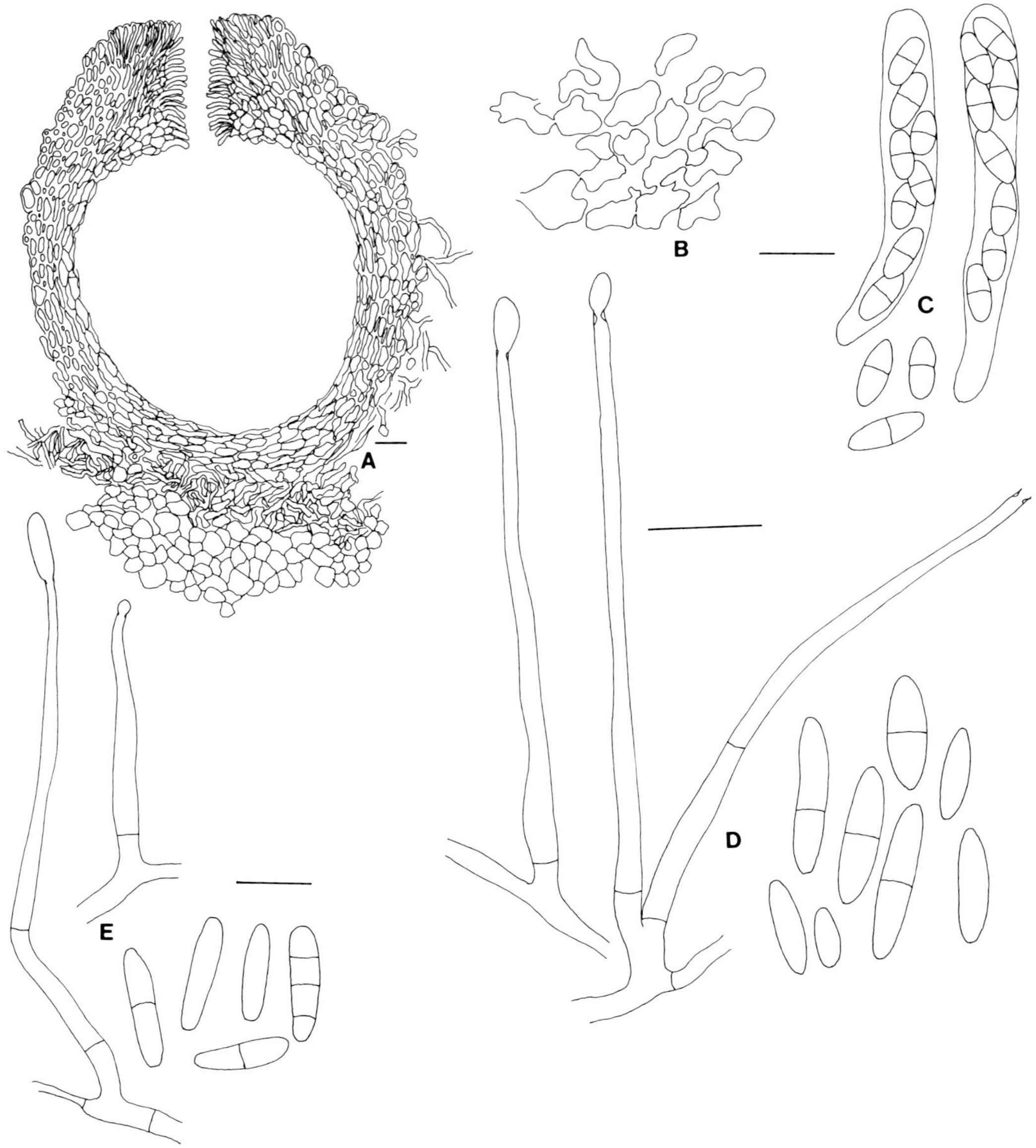

FIG. 9. *Nectriopsis lasiodermopsis.* **A.** Section through a mature perithecium (*Doi 6761*). **B.** Cells at the surface of the perithecial wall (*Doi 6766*). **C.** Asci and ascospores (*Doi 6766*). **D.** Conidiophores and conidia from New Zealand (holotype). **E.** Conidiophores and conidia from Australia (*Doi 6761*). Lines = 10 μm.

hyphae, macronematous, mononematous, unbranched, smooth, 1–2-septate, (58–)75–132 (–160) μm long × 2–3 μm wide at base, monophialidic. Phialides integrated, terminal, straight, smooth, (13–)25–66(–80) μm long, monoblastic, tip slightly thickened, flared or not, (1.5–)2(–2.5) μm. Conidia cylindrical to naviculate and narrower at tip than at base, with a broad, protuberant basal abscission scar:

0-septate: (10–)11–14.5(–16) × (2.5–)2.7–3.7(–4) μm
1-septate: (14–)14.8–21(–22) × 3–4 μm
2-septate: (17–)18–23(–25) × 3–4 μm
3-septate: (17–)20.5–26 × (3–)3.5–5 μm;

held in a single, terminal, colorless drop of slime.

Habitat. On perithecia of *Nectria* spp.

Known Distribution. New Zealand (Auckland).

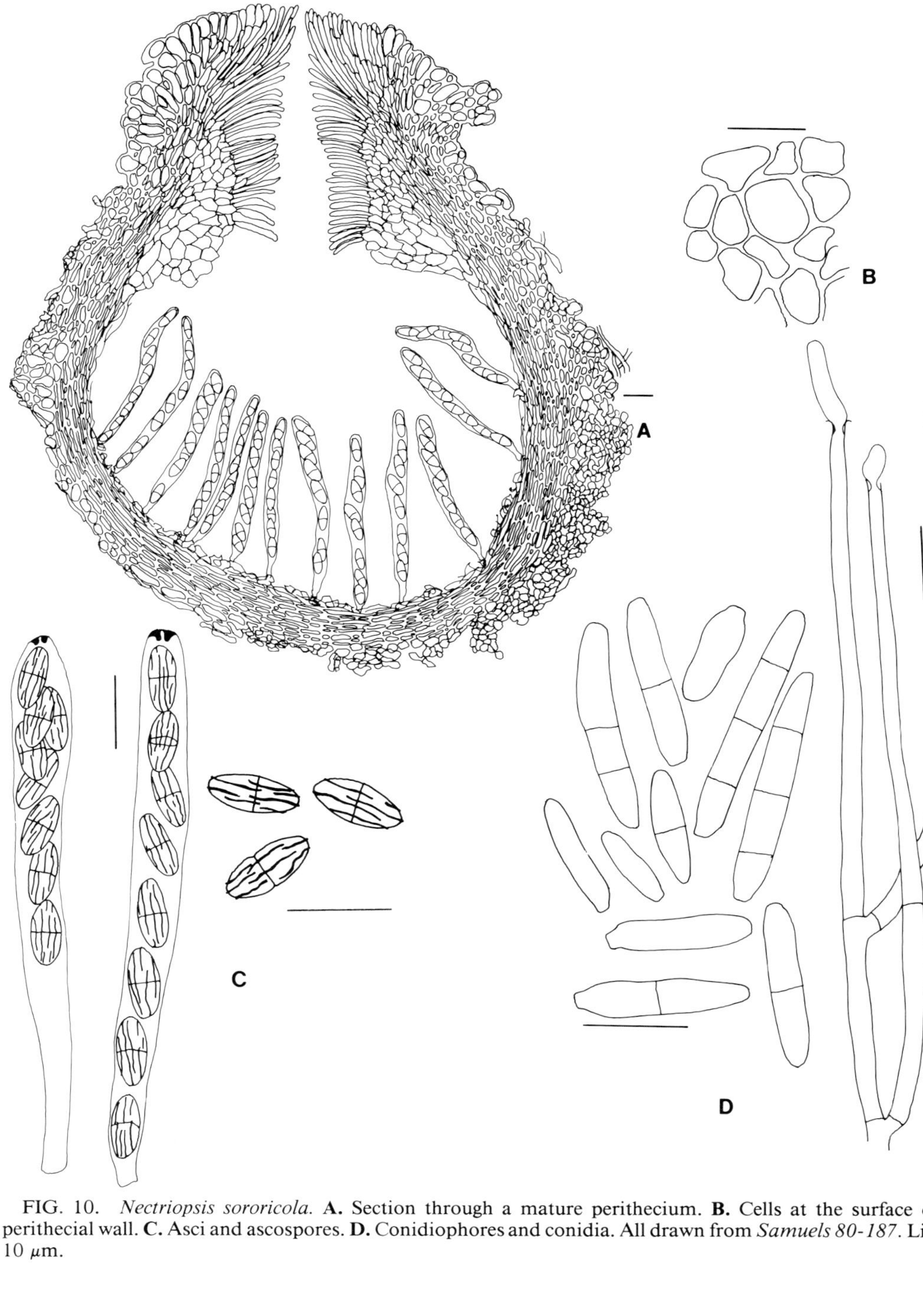

FIG. 10. *Nectriopsis sororicola.* **A.** Section through a mature perithecium. **B.** Cells at the surface of the perithecial wall. **C.** Asci and ascospores. **D.** Conidiophores and conidia. All drawn from *Samuels 80-187*. Lines = 10 μm.

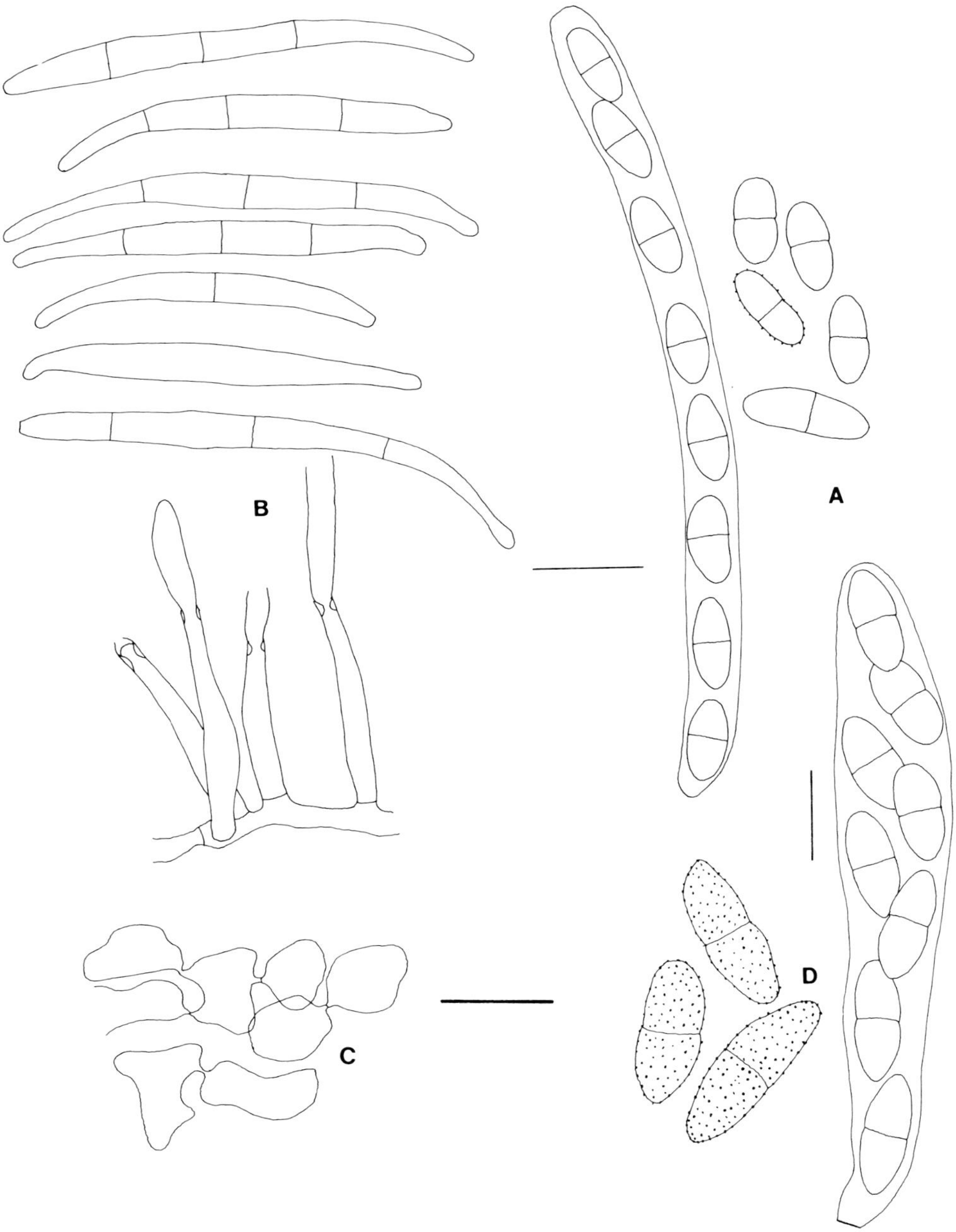

FIG. 11. A, B. *Nectriopsis sibicola.* **A.** Ascus and ascospores. **B.** Conidiophores and conidia. Both drawn from the holotype. C, D. *Nectriopsis hyperbiota.* **C.** Cells at the surface of the perithecial wall. **D.** Ascus and ascospores. Both drawn from PDD 25002. Lines = 10 μm.

Etymology of the Specific Epithet. Refers to parasitism on a species of a sister genus of the Hypocreales.

Holotype. New Zealand. North Island. Auckland: Waitemata City, Titirangi, Titirangi Beach Reserve, on perithecia of *Nectria discophora* (Mont.) Mont., *Samuels, Johnston & Rattray,* 10 Dec 1980 (PDD 41423).

Additional Specimen Examined. NEW ZEALAND. North Island. AUCKLAND: Waitemata City, Waitakere Ranges, off Mountain Rd., on *Nectria tawa* Dingley (PDD 18527), *Dingley,* 16 Sep 1958 (PDD 51331, NY).

Notes. Cultures derived from the holotype are no longer viable. The ring in the apex of asci of PDD 51331 is obscure, far less obvious than it is in asci of the holotype.

Nectriopsis sororicola superficially resembles *N. puiggarii* and *Nectria brunneostriata* in forming small yellow perithecia directly on the perithecia of other pyrenomycetes. *Nectriopsis albofulta* and *N. epinectria* also have striate ascospores but perithecia of these species form within mycelium on other ascomycetes. Ascospores of *Nectria brunneostriata* and *N. epinectria* are distinctly brown.

The anamorph of *N. sororicola* cannot be unequivocally placed into any described anamorph genus. The phragmosporous conidia borne on phialides indicate *Cylindrocarpon* or *Cephalosporiopsis.* Because *N. sororicola* is unrelated to any of the species of *Nectria* that have typical *Cylindrocarpon* anamorphs (Booth, 1966), I do not refer this anamorph to that genus. *Cephalosporiopsis* and *Acremonium,* as described, are distinguished through conidial septation; conidia of species of the former are once septate whereas conidia of species of the latter are aseptate. Rossman (1983) emended the concept of *Cephalosporiopsis* to include the species with phragmosporous conidia that are anamorphs of *Nectria arenula* (Berk. & Broome) Berk. and its relatives (Rossman, 1983; Samuels, 1978). *Nectriopsis sororicola* is not closely related to these fungi. The closest relatives of *N. sororicola* are species of *Nectriopsis,* and most of the known anamorphs of *Nectriopsis* are species of *Acremonium.* In the belief that anamorphs of related teleomorphs are themselves related, in spite of morphological differences, and in the understanding that *Acremonium* as currently constituted is biologically heterogeneous, I refer the anamorph of *N. sororicola* to *Acremonium.*

27. Nectriopsis sibicola Samuels, sp. nov. (Fig. 11A, B).

Nectriopsis sororicolae Samuels similis, sed ascosporis brevioribus et angustrioribus, (7–)7.8–9.6(–10) × 4–4.5(–5) μm, et statu anamorphicus, *Acremonium* simulanti, differt.

Holotypus. New Zealand (PDD 46339).

Anamorph. *Acremonium*-like.

Mycelium not evident or at most thin, white and spreading. Perithecia cespitose in groups of a few, superficial, nonstromatic but difficult to remove from the substrate, pyriform, 230–250 × 185–220 μm, with an acute papilla, smooth, collapsing by lateral pinching or not collapsing when dry; yellow, not changing color in 3% KOH.

Cells at surface of perithecial wall forming *textura epidermoidea,* often obscured by 2–3 μm wide hyphae. Perithecial wall 10–20 μm wide, comprising a single region of small, indistinct cells; perithecial papilla formed of a palisade of hyphal elements that become progressively narrower and more thin walled toward the ostiolar canal and there merging with the periphyses.

Asci clavate, 60–85 × 7–9 μm, apex simple; 8-spored, ascospores 1-seriate, completely filling each ascus. Ascospores ellipsoidal, (7–)7.8–9.6 (–10) × 4–4.5(–5) μm, equally 2-celled, not constricted at the septum, smooth to finely spinulose, colorless.

Characteristics in Culture. Colonies grown 14 days at 15–18°C diffuse daylight on CMD, PDA, OA 1 mm diam., white, aerial mycelium lacking, colony comprising tightly compacted hyphae and conidiophores. Colonies grown four days at 20°C, 12 hours dark/12 hours near ultraviolet + cool white fluorescent light becoming pale yellow on CMD, PDA and OA. Conidiophores arising from 2–3 μm wide hyphae, macronematous, mononematous, straight, smooth, basally uniseptate, monophialidic, 18–28 × 2–2.5 μm at base; tip 1–1.5 μm wide, thickened, not flared. Conidia (29–)35.5–53(–67) × 3.5–4 μm, 0–5-septate, cylindrical but with tip elongated, narrowed and often hooked; base truncate and often with a poorly defined, often foot-like protrusion; produced in basipetal succession and held in white slime.

Habitat. On perithecia of *Nectria* sp.

Known Distribution. New Zealand (South Island).

Etymology of the Specific Epithet. Refers to parasitism on a species of a sibling genus of the Hypocreales.

Holotype. New Zealand. South Island. Buller: 7 km S of Reefton, Tawhai Forest, on *Nectria* sp. on *Coprosma foetidissima* Forst., *Samuels, Matsushima & Rossman,* 22 May 1983 (PDD 46339).

Notes. *Nectriopsis sibicola* and *N. sororicola* are most easily distinguished through their anamorphs; their teleomorphs are very similar. On the basis of one collection of *N. sibicola* and two collections of *N. sororicola,* ascospores of *N. sibicola* are somewhat shorter and broader than those of *N. sororicola.*

Like the anamorph of *N. sororicola,* the anamorph of *N. sibicola* cannot be unequivocally placed in any anamorph genus. Conidia of *N. sibicola* are somewhat *Fusarium*-like, but *N. sibicola* is not closely related to any of the known teleomorphs of *Fusarium* (Gerlach & Nirenberg, 1982). Because the closest relatives of *N. sibicola* are *Nectriopsis* species with *Acremonium* anamorphs, I have followed the same reasoning as that applied to *N. sororicola,* placing the anamorph of *N. sibicola* in *Acremonium.*

28. Nectriopsis hyperbiota Samuels, sp. nov. (Fig. 11C, D).

Nectriopsis perpusillae (Mont.) Samuels similis, sed ascosporis majoribus, (12–)13.7–17.2 (–21) × (5–)5.7–7(–8) μm.

Status anamorphicus ignotis.
Holotypus. New Zealand (PDD 7509).
Anamorph. None known.

Mycelium not evident. Perithecia cespitose in groups of several, superficial, nonstromatic but difficult to remove from the substrate; pyriform, 150–250 × (115–)125–170(–220) μm, with an acute papilla, smooth, collapsing by lateral pinching when dry; yellow-orange, not changing color in 3% KOH. Cells at the surface of the perithecial wall forming *textura epidermoidea,* walls unevenly thickened, lumina of adjacent cells joined by fine pores. Perithecial wall ca. 20 μm wide, comprising a single region of small, indistinct cells; papilla formed of a palisade of hyphal elements that become progressively narrower and more thin-walled toward the ostiolar canal and there merging with the periphyses.

Asci clavate, (62–)68–83(–90) × (10–)11.7–14.7(–16) μm, apex simple; 8-spored, ascospores biserate, completely filling each ascus. Ascospores ellipsoidal to fusiform, (12–)13.7–17.2 (–21) × (5–)5.7–7(–8) μm, equally 2-celled, not constricted at the septum, pronouncedly spinulose, pale brown.

Habitat. On perithecia of *Nectria pseudotrichia* Berk. & Curt.

Known Distribution. New Zealand (North Island).

Etymology of the Specific Epithet. Refers to the hyperparasitic habit.

Holotype. New Zealand. North Island. Auckland: Waitemata City, Waitakere Ranges, off Anawhata Rd., on *Nectria pseudotrichia* on *Dysoxylum spectabile* (Forst. f.) Hook. f., *Dingley,* Apr 1948 (PDD 7509; isotype NY).

Additional Collections Examined. NEW ZEALAND. North Island. AUCKLAND: Waitemata City, Waitakere Ranges, Huia, on *Nectria pseudotrichia, Dingley,* 7 Feb 1954 (PDD 18210, NY); Waitakere Ranges, Waitakere Dam, on *Nectria pseudotrichia, Dingley (6651),* 9 Feb 1966 (PDD 25002, NY).

29. Nectriopsis cephalosporii (Hansford) Samuels, comb. nov. (Fig. 12A–C).

≡*Calonectria cephalosporii* Hansford, Mycol. Pap. **15:** 117. 3 Apr 1946.

Anamorph. *Acremonium* sp.

Mycelium white, cottony, conidiogenous. Conidiophores arising from aerial mycelium, sometimes growing out of hyphae of the hose fungus, macronematous, mononematous, unbranched, multiseptate, monophialidic. Phialides 30–40 μm long × 3–5 μm wide at base, tapering uniformly to 1–1.5 μm wide at the tip, thick-walled, tip with slight periclinal thickening, collarette not flared. Conidia oblong to ellipsoidal, 7.5–9(–10) ×

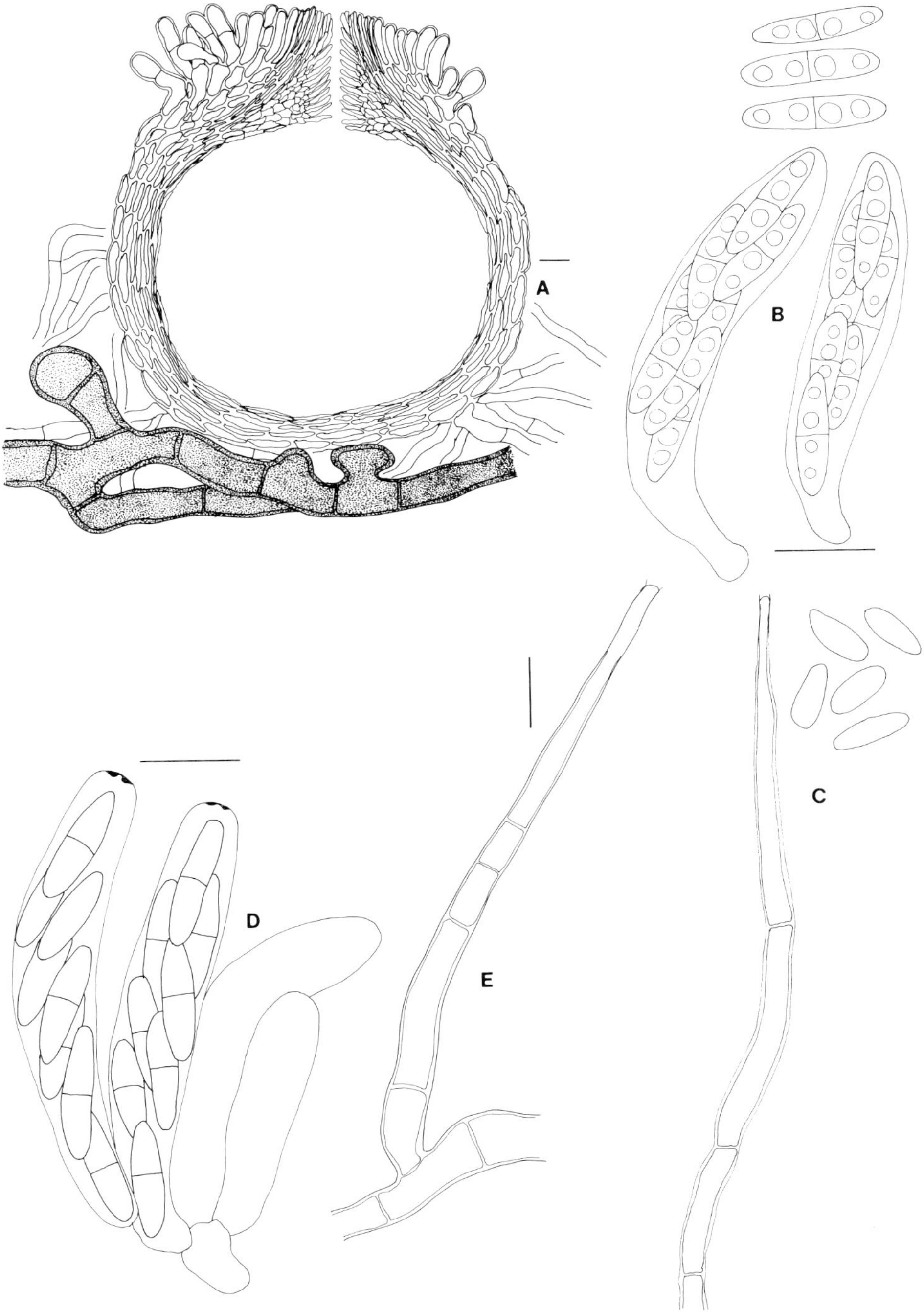

FIG. 12. A–C. *Nectriopsis cephalosporii.* **A.** Section through a mature perithecium (*Thaxter*). **B.** Asci and ascospores (type). **C.** *Acremonium* conidiophore and conidia (type). D, E. *Nectriopsis sensitiva.* **D.** Two mature asci and two immature asci. **E.** Conidiophore. D, E drawn from the holotype. Lines = 10 μm.

(2–)2.5–3(–3.5) μm, lacking a visible basal abscission scar, unicellular, biguttulate, colorless, smooth, held in a drop of colorless liquid at the tip of each phialide.

Perithecia scattered, seated on white mycelium, nonstromatic, easily removed from the substrate, globose, 100–160(–220) μm diam., nonpapillate, smooth; hairs arising from perithecial apex, hairs (10–)15–25(–35) μm long × 3–5 μm wide, flexuous, septate, unbranched, tip plane or swollen, wall 0.5–1 μm thick. Perithecia collapsing by lateral pinching when dry, yellow, not changing color in 3% KOH.

Cells at surface of perithecial wall obscured by hyphae. Perithecial wall 8–10 μm wide, comprising a single region of nondescript, small cells; perithecial apex of diverging hyphal elements, many of which grow outwardly to form the perithecial hairs and merging with the periphyses within.

Asci clavate, (30–)33–46(–53) × (6–)7–9(–10) μm, apex simple, 8-spored, ascospores biseriate, completely filling each ascus. Ascospores ellipsoidal to fusiform, (8.5–)11.4–15.5(–18) × 2.3–4 μm, equally 2-celled, not constricted at the septum, smooth, colorless.

Habitat. On members of the Meliolaceae on leaves.

Known Distribution. Neotropics, Africa, East Indies.

Holotype. UGANDA. Kawanda, on *Meliola markhamiae* Hansford & F. L. Stevens, on leaves of *Markhamia platycalyx* Sprague *in* Hook. (Bignoniaceae), *Hansford 2554b,* Aug 1939 (IMI 4680b!).

Additional Specimens Examined. PUERTO RICO. Cerro de Las Pinas, nr. Las Cruces, 600–700 m, on *Meliola* sp. on *Casearia arborea* Urb., *Britton, Matz & Chardon 7019,* 29 Mar 1922 (NY). TRINIDAD. Port of Spain, Maraval Valley, on *Meliola* sp. on *Piper* sp., *R. Thaxter,* 1912–1913 (FH). PERU. Oromina Rubber Nursery, on *Meliola* sp. on *Hevea brasiliensis* Muell.-Arg., *R. Russell 8004,* Jul 1942 (BPI). BRAZIL. RORAIMA: Along the Boa Vista–Dormida Rd. at a point ca. 23 km from the intersection of the Boa Vista–Bonfim Rd. and at a point ca. 5 km from Boa Vista, on *Meliola* sp. on live dicot. leaves, *Dumont (BR 898, BR 906), Hosford, Samuels, Buck, Araujo, Sousa & Bernardi,* 4 Dec 1977 (NY). CHILE. LOS LAGOS: Corral, on *Meliola* sp., *R. Thaxter,* 1905 (RH).

Notes. *Nectriopsis cephalosporii* is most closely related to *N. sensitiva,* from which it differs in having a simple ascal apex and in having perithecial hairs.

Nectriopsis cephalosporii has never been grown in pure culture but perithecia of the species are nearly constantly associated with the unusual *Acremonium* anamorph described above. For an illustration and a more complete discussion of this anamorph see Gams (1971).

The collections cited herewith are the first records of this fungus for the New World.

30. Nectriopsis sensitiva (Rehm) Samuels, comb. nov. (Fig. 12D, E).

≡*Nectria sensitiva* Rehm, Hedwigia **39:** 222. 10 Aug 1900.

Anamorph. *Acremonium* sp.

Mycelium white, cottony, conidiogenous; hyphae ca. 5 μm wide, septate, branching, smooth. Conidiophores arising from aerial hyphae, macronematous, mononematous, monophialidic, 80–100 μm long × ca. 7 μm wide at base, septate, thick-walled, straight, smooth; tip of phialide lacking visible periclinal thickening, collarette not flared. Conidia not seen.

Perithecia scattered, seated on white mycelium, nonstromatic, easily removed from the substrate, subglobose, ca. 125 μm diam., nonpapillate, smooth, lacking hairs, collapsing by lateral pinching when dry; pale yellow, not changing color in 3% KOH.

Cells at surface of the perithecial wall pseudoparenchymatous, 7–10 μm across, walls ca. 1 μm thick. Perithecia thin-walled, wall comprising a single region of nondescript, small cells.

Asci clavate, 45–60 × 7–12 μm, apex with a small ring; 8-spored, ascospores biseriate, completely filling each ascus. Ascospores ellipsoidal, (13.5–)14–16(–17) × 2–3.5(–4) μm, equally 2-celled, not constricted at the septum, smooth, colorless.

Habitat. On *Meliola* sp.

Known Distribution. Brazil, known only from the type.

Holotype. Brazil. Estado do Rio de Janeiro, Tijuca, on leaf of Mimosaceae, *Ule 2274,* May 1897 (K!).

Notes. *Nectriopsis sensitiva* is very close to *N. cephalosporii,* differing in having a ring in the ascal apex and in the lack of perithecial hairs. Both species are associated with *Acremonium* anamorphs that are unusual in having thick-walled conidiophores and phialides. The conidiophores of these species are very similar but

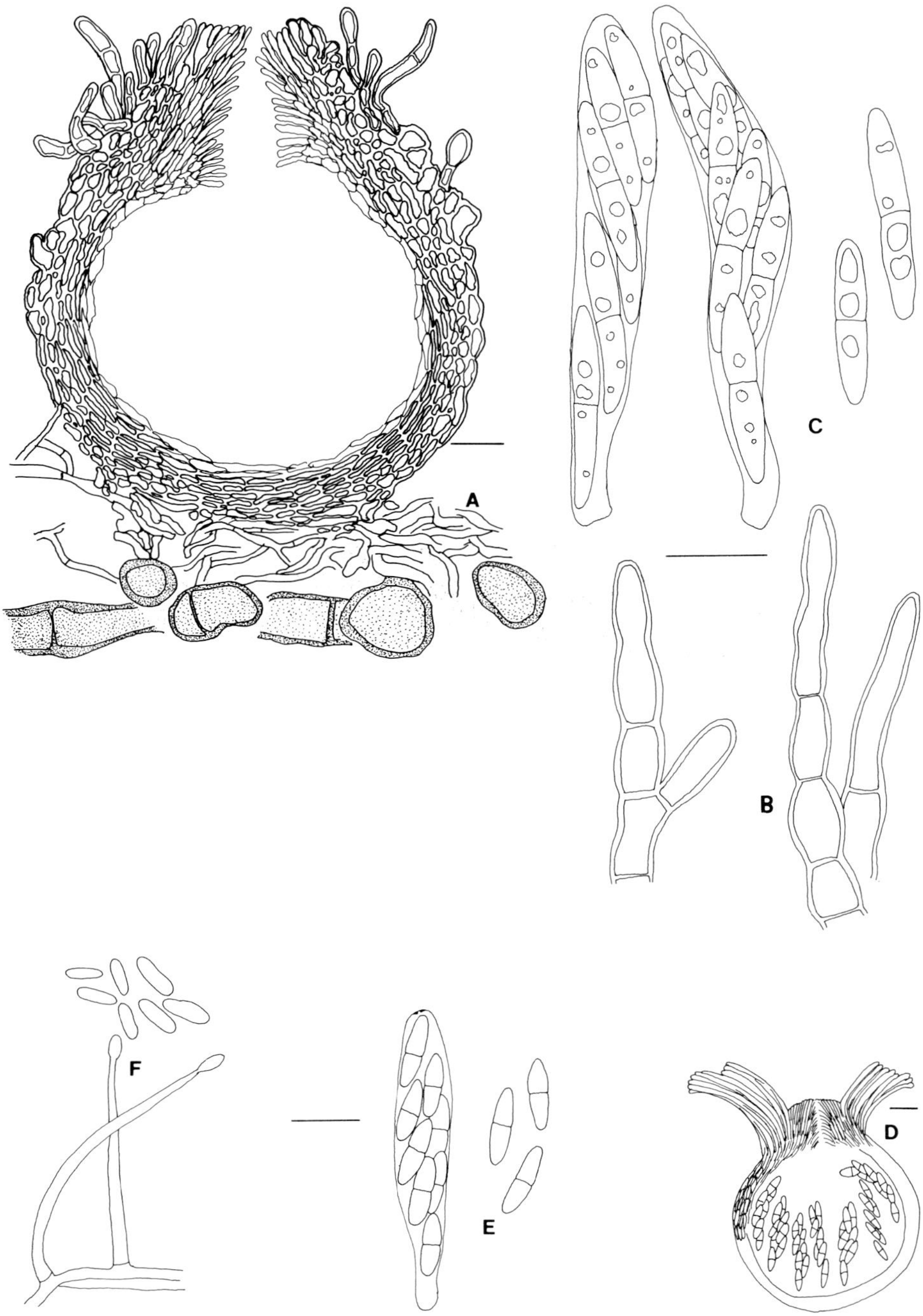

FIG. 13. A–C. *Nectriopsis oidioides.* **A.** Section through a mature perithecium. **B.** Perithecial hairs. **C.** Asci and ascospores. Drawn from the *Thaxter* collection. D–F. *Nectriopsis porinicola.* **D.** Diagrammatic section through a mature perithecium. **E.** Ascus and ascospores. **F.** Conidiophores and conidia. Drawn from the holotype. Lines = 10 μm.

because no conidia remain on the type and only known specimen (herewith, and Hansford, 1946) of *N. sensitiva,* I am unable to distinguish between the two anamorph species.

31. Nectriopsis oidioides (Spegazzini) Samuels, comb. nov. (Fig. 13A–C).

≡*Nectria oidioides* Spegazzini, Bol. Acad. Nac. Cienc. Córdoba **11:** 524. Jul 1889.
≡*Cucurbitaria oidioides* (Spegazzini) O. Kuntze, Rev. Gen. Pl. **3(2):** 461. 28 Sep 1898.

Anamorph. None known.

Mycelium white, restricted to colonies of a meliolaceous fungus. Perithecia solitary, superficial, seated in white mycelium, subglobose, 110–150 μm diam., smooth, collapsing by lateral pinching when dry; yellow, not changing color in 3% KOH, with hairs growing from the upper half of the perithecial wall; hairs cylindrical with ± clavate ends, 25–30 μm long × 4–7 μm wide at the tips.

Cells at surface of perithecial wall forming *textura epidermoidea.* Perithecial wall 5–10 μm wide, comprising a single region of small cells 3–5 × 2 μm, walls ca. 1 μm thick. Perithecial apex formed of 2 μm wide hyphal elements.

Asci narrowly clavate, (45–)50–65(–70) × 9–12 μm, apex simple; 8-spored, ascospores partially to completely biseriate, completely filling each ascus. Ascospores narrowly fusiform to naviculate, (14–)16.8–22.2(–27) × 3–4 μm, equally 2-celled, not constricted or slightly constricted at the septum, finely striate or appearing smooth, colorless.

Habitat. On meliolaceous fungi on living leaves of dicotyledonous plants.

Known Distribution. Brazil, Trinidad.

Holotype. Brazil. São Paulo: Apiahy, on Sapindaceae, *Puiggari 1507,* iv.1881 (LPS!).

Additional Specimen Examined. TRINIDAD. Port of Spain, Maraval Valley, on meliolaceous fungus, *Thaxter,* 1912–1913 (FH).

Notes. *Nectriopsis oidioides* is distinguished among the meliolicolous species (Samuels, 1976a) by its large ascospores and by the perithecial hairs.

The specimen reported by Batista et al. (1967) as *Nectria meliolicola* F. L. Stevens is probably *Nectriopsis oidioides.*

32. Nectriopsis leucorrhodina (Montagne) Samuels, comb. nov.

≡*Peziza leucorrhodina* Montagne *in* Sagra, Hist. Cuba Pl. Cell. p. 360. 1842.
=*Nectria puberula* Spegazzini var. *microspora* Batista & Nascimento *in* Batista, Nascimento & Vital, Inst. Micol. Recife Publ. **33:** 5. 1956.

Anamorph. *Acremonium* sp.

Mycelium white, restricted to black colonies of host fungus. Perithecia superficial, nonstromatic, globose (160–)166–245(–270) μm diam., yellow, not changing color in 3% KOH. Asci subfusiform, 35–45(–50) × 5–11 μm, apex simple. Ascospores elliptic-fusiform to fusiform, (11–)12.4–16.3(–19) × 3–4 μm, equally 2-celled, striate, colorless.

Habitat. On meliolaceous fungi on living leaves.

Note. See Samuels (1976a, as *Nectria leucorrhodina*) for illustrations, a complete description, and additional synonymy.

33. Nectriopsis guarapiensis (Spegazzini) Samuels, comb. nov.

≡*Calonectria guarapiensis* Spegazzini, Anales Soc. Ci. Argen. **19:** 41. Jan 1885.
≡*Nectria microleuca* Rossman, Mycotaxon **8:** 515. 21 Apr 1979.
=*Nectria bakeri* Rehm, Ann. Mycol. **6:** 319. 10 Oct 1908.
=*Nectria perpusilla* Saccardo, Ann. Mycol. **11:** 515. 31 Dec 1913, *non* (Mont.) Mont. 1856.

Anamorph. None known.

Mycelium white, restricted to black hyphae of host fungus. Perithecia pyriform, 115 × 105 μm, yellow, not changing color in 3% KOH, with cylindrical, septate hairs arising from the upper half of the perithecial wall. Asci clavate, 42–55 × 5–6 μm, apex with a ring. Ascospores ellipsoidal to fusiform-ellipsoidal, 8–10.5 × 2.5–3 μm, equally 2-celled, finely spinulose.

Habitat. On meliolaceous fungi on living leaves.

Note. For additional synonymy see Rossman (1979); for illustrations and a complete description see Samuels (1976a, as *Nectria bakeri* Rehm).

34. Nectriopsis pipericola (Hennings) Samuels, comb. nov.

≡*Nectria pipericola* Hennings, Hedwigia **43:** 244. 12 Jun 1904.

Anamorph. None known.

Mycelium white, restricted to black colonies of host fungus. Perithecia globose to subglobose, (100–)110–214(–250) × 90–100(–250) μm, yellow, not changing color in 3% KOH, with cylindrical to napiform hairs, 15–20 μm long × 2–3 μm wide at base × 5–15 μm wide at tip. Asci clavate, 21–17 × 6–9 μm, apex simple. Ascospores oblong-ellipsoidal, 9–11 × 3–4 μm, smooth, colorless.

Habitat. On meliolaceous fungi on living leaves.

Note. For illustrations and a complete description see Samuels (1976a, as *Nectria pipericola*).

35. Nectriopsis porinicola Samuels, sp. nov. (Fig. 13D–F).

Perithecia pyriformia ±80 μm alta et ±70 μm lata, ad apicem acuta, pallide flavida, fere alba; setae triangulares 40 μm usque longae, ex hyphis 2–3 agglutinatis compositae, e perithecii apice orientes. Asci clavati 30–55 × 6.5–8 μm, ad apicem annulo refractili instructi. Ascosporae anguste ellipsoideae vel subcylindricae, (9–)9.5–10.7(–11.5) × 2.2–3(–3.3) μm, laeves, hyalinae.

Status anamorphicus *Acremonium* sp.
Holotypus. Indonesia (BO: *Samuels 2155*).
Anamorph. *Acremonium* sp.

Mycelium not evident. Perithecia solitary, scattered, superficial, nonstromatic, easily removed from the substrate, pyriform, ca. 80 μm high × ca. 70 μm wide, with an acute papilla, collapsing by lateral pinching when dry; pale yellow, nearly white, not changing color in 3% KOH; with fasciculate, triangular, 40 μm long hairs radiating outwardly from the perithecial apex, each hair comprising 2–3, agglutinated, septate, 2–3 μm wide hyphae with walls ca. 1 μm thick.

Cells at the surface of the perithecial wall forming thin-walled *textura epidermoidea*. Perithecial wall 5–7 μm wide, comprising a single region of small, indistinct cells; perithecial papilla formed of hyphal elements, elements becoming progressively narrower and merging with the periphyses within.

Asci clavate, 30–55 × 6.5–8 μm, apex with a minute refractive ring; 8-spored, ascospores biseriate, completely filling each ascus. Ascospores narrowly ellipsoidal to nearly cylindrical, (9–)9.5–10.7(–11.5) × 2.2–3(–3.3) μm, equally 2-celled, not constricted at the septum, smooth, colorless.

Characteristics in Culture. Colonies grown 14 days at 20°C diffuse daylight on CMD 1 cm diam., flat, white, opaque, with scant aerial mycelium; conidiophores arising abundantly from the surface of the colony. Conidiophores macronematous, mononematous, arising from ca. 2 μm wide hyphae, straight, smooth, with a single, basal septum, 22–35 μm long × ca. 2 μm at base, monophialidic; phialide terminal, integrated, 1 μm wide at tip, at most slightly thickened, not flared. Conidia oblong, 4.5–7(–9) × 1.7–2.2 μm, without a conspicuous basal abscission scar, unicellular, colorless, held in a drop of colorless liquid.

Habitat. On ascomata of the lichen *Porina epiphylla* (Fee) Fee *s. lat.*

Known Distribution. Indonesia (North Sulawesi), known only from the type collection.

Etymology of the Specific Epithet. Refers to the substrate of the type collection, *Porina epiphylla.*

Holotype. Indonesia. North Sulawesi: Dumoga-Bone National Park, between 'Maddison's Camp' (0°36′N, 123°50′E) and '1440 Camp' (0°37′N, 123°51′E), 980–1100 m, on *Porina epiphylla, Samuels 2155,* 5 Oct 1985 (BO; isotype NY).

36. Nectriopsis hypocrellicola (Hennings) Samuels, comb. nov. (Fig. 14A, B).

≡*Nectria hypocrellicola* Hennings, Hedwigia **41:** 4. 5 Mar 1902.

Anamorph. None known.

Mycelium not evident. Perithecia cespitose in a continuous layer of several, pyriform, 175–260 μm high × 157–220 μm wide, apex rounded; perithecia smooth, collapsing by lateral pinching when dry, yellow, not changing color in 3% KOH.

Cells at surface of the perithecial wall angular, 7–10 μm in greatest dimension, walls 0.5–1 μm thick. Perithecial wall 25–30 μm wide, cells angular to nearly elliptical in outline, 7–10 μm in greatest dimension, walls 1–1.5 μm thick; ostiolar area formed of hyphal elements with rounded ends, elements becoming progressively thinner toward the ostiolar canal and there merging with the periphyses.

Asci cylindrical, (69–)72–84(–85) × 6–7 μm, apex simple; 8-spored, ascospores uniseriate, completely filling each ascus. Ascospores ellipsoidal, (8–)9–11 × 4–5 μm, equally 2-celled, not or slightly constricted at the septum, smooth to slightly tuberculate, pale yellow (100% lactic acid).

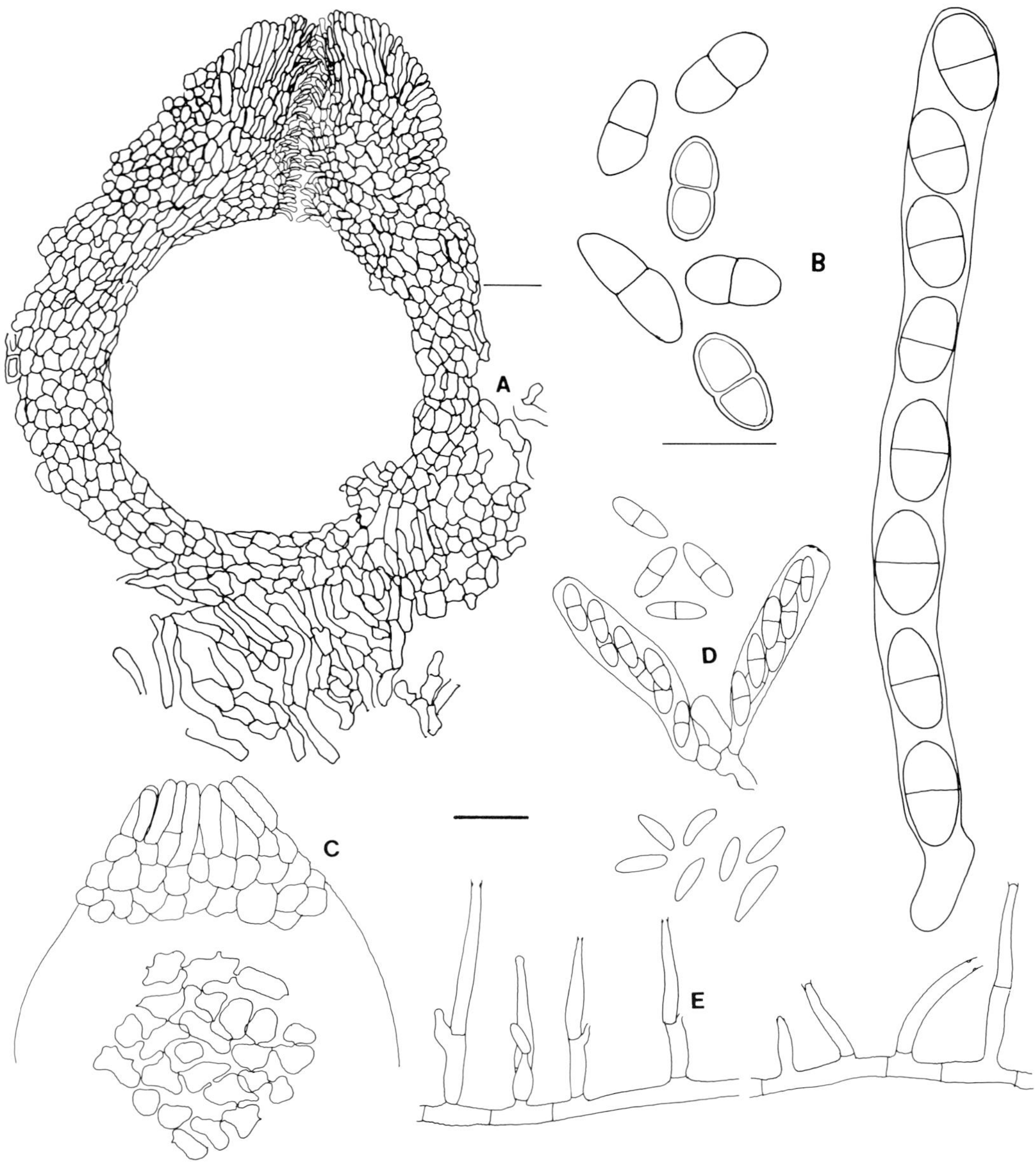

FIG. 14. A–B. *Nectriopsis hypocrellicola.* **A.** Section through a mature perithecium. **B.** Ascus and ascospores. A, B drawn from the holotype. C–E. *Nectriopsis microthecia.* **C.** Surface of a mature perithecium showing the perithecial apex and cells at the surface of the perithecial wall. **D.** Asci and ascospores. **E.** Conidiophores and conidia. C–E drawn from the holotype. Lines = 10 μm.

Habitat. On black stroma on dead bamboo culm.

Known Distribution. Brazil, known only from the type.

Holotype. Brazil. Santa Catarina: Blumenau, in *Bambusa, Moeller,* 5.1893 (S!).

37. Nectriopsis squamulosa (Ellis) Samuels, comb. nov. (Fig. 15A–C).

≡*Nectria squamulosa* Ellis, Bull. Torrey Club **9:** 20. Feb 1882.

≡*Dialonectria squamulosa* (Ellis) Cooke, Grevillea **12:** 110. Jun 1884.

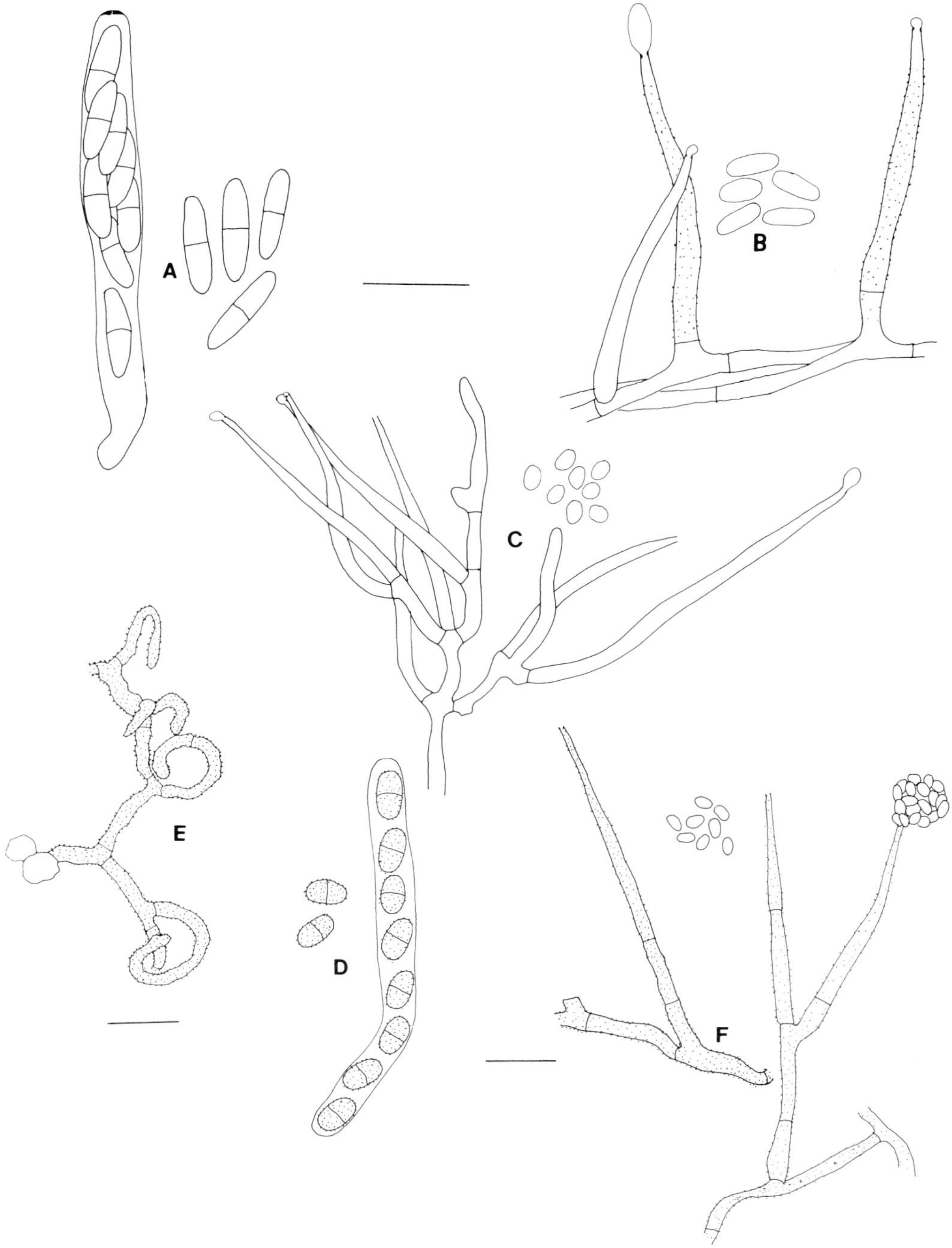

FIG. 15. A–C. *Nectriopsis squamulosa.* **A.** Ascus and ascospores (*GJS 74-111*). **B.** Conidiophores and conidia from culture (*GJS 74-111*). **C.** Conidiophores and conidia from nature (PDD 36421). D–F. *Nectriopsis oropensoides.* **D.** Ascus and ascospores. **E.** Hyphae from surface of perithecium. **F.** Conidiophores and conidia from nature. D–F drawn from *GJS 3982.* Lines = 10 μm.

=*Byssonectria chrysocoma* Cooke & Harkness, Grevillea **12:** 110. Jun 1884.
≡*Nectria chrysocoma* (Cooke & Harkness) Berlese & Voglino, Syll. Fung. Add. 204. 31 Dec 1886.
≡*Cucurbitaria chrysocoma* (Cooke & Harkness) O. Kuntze, Rev. Gen. Pl. **3(2):** 460. 28 Sep. 1898.

Anamorph. *Acremonium* sp.

Mycelium white to pale golden, scant, spreading; hyphae septate, branched, thin-walled, smooth, 2–3 μm wide. Conidiophores arising in the mycelium, macronematous, mononematous, 30–40 μm long × 1.5 μm wide at base, tapering to 1 μm wide at tip, straight, smooth or with yellow granulation at the base, colorless, monophialidic; tip of phialide not visibly thickened, not flared. Conidia ellipsoidal to subglobose, (2–)2.6–3.4(–4) × 2–2.5 μm, lacking a visible basal abscission scar, unicellular, colorless, held in a colorless drop of liquid at the tip of each phialide.

Perithecia gregarious, seated on the mycelium, nonstromatic, easily removed from the substrate, globose to broadly pyriform, (100–)116–160(–175) × (100–)111–147(–175) μm, nonpapillate, becoming cupulate, collapsing by lateral pinching or not collapsing when dry; smooth, wall covered by thin-walled, golden-yellow to white hyphae with few to many free ends; yellow to orange-yellow, either not changing color in 3% KOH or becoming roseous in 3% KOH.

Cells at surface of perithecial wall obscured by hyphae. Perithecial wall 10–15 μm wide, comprising a single region of small cells; perithecial apex not anatomically distinct from the lateral perithecial wall, ostiolar canal periphysate.

Asci broadly cylindrical to narrowlly clavate, (27–)33–49(–65) × 5–6.7(–8) μm, apex with a refractive apical ring; 8-spored, ascospores partially to completely biseriate, completely filling each ascus. Ascospores oblong to narrowly ellipsoidal, (6–)7.5–10.3(–14) × 1.5–2.3(–3) μm, ends obtuse, equally 2-celled, not constricted at the septum, smooth, colorless.

Characteristics in Culture. Colonies grown two weeks at 22°C, diffuse daylight. CMD, OA: flat, felty, opaque, pink, with concentric rings of conidial production; colony reverse pink; colonies on CMD with short aerial hyphae, colonies on OA with well developed, hyphal, rope-like strands, surface of colony on OA sulcate and crustose. Conidial production abundant on CMD and OA; conidiophores arising from surface of colony, aerial hyphae and hyphal ropes from 2 μm wide hyphae. Conidiophores macronematous, mononematous, unbranched or once branched, 20–35 μm long × 2 μm wide at base, tapering uniformly to 1 μm wide at tip, with a single, basal septum; smooth or roughened at the base, colorless. Conidia ellipsoidal to subglobose, (3–)3.3–5.3(–7) × 1.5–2 μm, basal abscission scar not visible, unicellular, colorless, held in a pale salmon-colored drop of liquid at the tip of each phialide.

Habitat. On herbaceous debris including palms, Pandanaceae and the tree fern *Cyathea dealbata* (Forst. f.) Swartz.

Known Distribution. Colombia, France, New Zealand, USA (California, Massachusetts, New Jersey, New York).

Holotype. USA. New Jersey: Newfield, on fallen limb, *Ellis 656b,* Nov 22, 1881 (NY!).

Additional Specimens Examined. USA. CALIFORNIA: On *Eucalyptus* sp., *Harkness 2321* (holotype of *Byssonectria chrysocoma,* NY). MASSACHUSETTS: Sharon, on rotting hardwood, *Piquet* (NY ex FH). NEW YORK: Adirondack Mts., on *Chondrioderma spumarioides* (Fries) Rost., *Rex,* Aug 1882 (NY, in type packet of *Nectria rexiana*). COLOMBIA. CUNDINAMARCA: Ca. 15 km from Bogotá, on the Bogotá–Choachi Rd., on *Puya* leaves, *Dumont (CO 5539), Carpenter & Sherwood,* 4 Aug 1976 (NY). FRANCE. PYRENÉES-ATLANTIQUES: Pau, on *Phyllostachys mitis* A. & C. Rivière, *Dr Travade comm. F. Candoussau,* 2 Apr 1983 (PDD 46476, NY); LANDES: Capbreton, Bord du Boudigau, on *Phyllostachys mitis, Candoussau 4862,* 9 Jun 1986 (Herb. Candoussau, NY). NEW ZEALAND. North Island. AUCKLAND: Waitemata City, Waitakere Ranges, Fairy Falls Track, on *Rhopalostylis sapida* Wendl. & Drude, *Samuels 75-148* (PDD 36421, NY); Fairy Falls Track, on *Cyathea dealbata, Samuels,* 22 May 1975 (PDD 36420); North Island. COROMANDEL: Thames Co., Coromandel Forest Park, vic. Thames, Kauaeranga Valley, on decaying leaf of *Freycinetia baueriana* Endl. subsp. *banksii* (A. Cunn.) Stone, *Dingley, Samuels (74-111) & Haydon,* 27 Aug 1974 (PDD 34054).

Notes. This concept of *N. squamulosa* may be too broad but I cannot see characters that are sufficiently consistent to recognize more than one taxon. The roseous perithecial coloration in 3% KOH was seen only in the French and New Zealand collections.

Ascospores of the type specimen of *Nectria squamulosa* are 7–9 × 1.5–2 μm, ascospores of the type collection of *Byssonectria chrysocoma* are 8–12 × 2–3 μm; perithecia of this collection are associated with a thin, golden-yellow mycelium.

The characteristics of the naturally formed

anamorph are taken from two New Zealand collections (PDD 36420, 36421). One of these (PDD 36420) was cultured; a second New Zealand isolate (PDD 34054) was also cultured. Conidia found in cultures of PDD 36420 were identical to those found on the specimen whereas conidia found in cultures of 34054 were longer [(4–)4.3–5.8(–7) × 1.5–2 μm] and ellipsoidal. Perithecia of PDD 34054 did not become roseous in 3% KOH. Conidia of the French specimens were identical to those found on the New Zealand specimens. Cultures of the New Zealand collections are no longer viable.

Nectriopsis squamulosa is most closely related to the myxomyceticolous species *N. candicans* and *N. oropensoides*.

38. Nectriopsis oropensoides (Rehm) Samuels, comb. nov. (Fig. 15D–F).

≡*Nectria oropensoides* Rehm *in* Tavel, Unters. Gesammt. Mykol. X Hefte, Ascom. **2:** 175. 1891.
≡*Cucurbitaria oropensoides* (Rehm) O. Kuntze, Rev. Gen. Pl. **3(2):** 461. 28 Sep 1898.
=*Nectria lactea* Ellis & Everhart, N. Amer. pyrenomyc. 110. 1892.

Anamorph. *Acremonium* sp.

Mycelium white, spreading, villose, with many erect hyphae and free hyphal ends; hyphae ca. 2 μm wide, septate, branched. Conidiophores arising in the aerial mycelium, macronematous, mononematous, 35–50 μm long, straight, smooth or spinulose, unbranched and monophialidic or branched once or twice, each branch terminating in a single phialide. Phialides 30–40 μm long × 2–3 μm wide at base, straight, smooth or spinulose, periclinal thickening not visible at tip, collarette not flared. Conidia globose to subglobose, (2–)2.7–3.5(–4) × (1.5–)1.8–3 μm, basal abscission scar not visible.

Perithecia solitary to gregarious, seated in white mycelium, superficial, nonstromatic, easily removed from the substrate, globose, (137–)190–210(–280) μm diam.; becoming cupulate when dry, amber-colored to white, not changing color in 3% KOH, wall covered by white hyphae, hyphae arising from surface of perithecial wall branched, septae, thin-walled, spinulose, 2–3 μm, with many free ends, ends often circinate.

Cells at the surface of the perithecial wall forming *textura epidermoidea.* Perithecial wall 7–10 μm wide, comprising a single region of small, nondescript cells; perithecial apex formed of narrow, ca. 2 μm wide hyphal elements; hyphal elements becoming progressively narrower and merging within with the periphyses.

Asci cylindrical, (37–)45–57(–68) × (4–)4.5–6(–7) μm, apex with a ring; 8-spored, ascospores uniseriate, completely filling each ascus. Ascospores oblong to broadly ellipsoidal, (4.5–)5.3–6.7(–8) × (2.5–)2.8–3.4(–4) μm, ends obtuse, equally 2-celled, not constricted at the septum, spinulose, colorless.

Characteristics in Culture. Colonies grown one week at 20°C diffuse daylight on CMD 1.5–2 cm diam., flat, transparent to opaque with aerial mycelium scant to dense, white; aerial hyphae short. Conidiophores arising throughout the colony, from 2 to 2.5 μm wide hyphae on surface of agar and in aerial mycelium; macronematous, mononematous, 25–65 μm long, colorless, smooth or spinulose, unbranched and monophialidic or once branched, each branch terminating in a single phialide. Phialides 20–45 μm long, tapering uniformly from 1.5–2.5 μm at base to 1 μm at tip, monoblastic, periclinal thickening of tip not visible, collarette not flared. Conidia subglobose to ellipsoidal, (2–)2.3–3.2(–4) × (1.3–)1.7–2.2(–3) μm, lacking a visible basal abscission scar, unicellular, colorless, smooth, held in a single drop of liquid at the tip of each phialide.

Habitat. On bark, herbaceous tissue, basidiocarps, rarely myxomycete sporangia.

Known Distribution. Canada, Colombia, France, French Guiana, Germany, Jamaica, Japan, New Zealand, USA (Florida, Indiana, Louisiana, New Jersey, New York, North Carolina, Ohio, South Carolina, Tennessee, Texas), Venezuela.

Holotype. Germany. On wood, *Tavel,* 10 Sep 1889 (S!).

Additional Specimens Examined. JAMAICA. ST. ANDREW PARISH: Holywell Forestry Camp, along the roadside, on *Pinus patula* Schiede & Deppe ex Schlecht., *Korf et al. (CUP-MJ 766),* 11 Jan 1971 (NY). COLOMBIA. VALLE: Gollondīna, nr. Cali, on decaying Aphyllophorales, *Doi,* 27, 28 Jun 1973 (NY, TNS 224601); second collection, same locality, on Thelephoraceae, *Doi,* 27, 28 Jun 1973 (NY, TNS 224599). VENEZUELA. AMAZONAS: Dpto. Río Negro, Cerro de la Neblina, along Río Mawarinuma, just outside Cañon Grande, vic. Neblina base camp, ca. 140 m; 00°50′N, 66°10′W, on bark, *Samuels 1435,* Apr–May 1984 (NY, VEN). FRENCH GUIANA. Km 16 on rd. between Sinnamary and St. Elie, 'ECEREX', ORSTOM research area, on *Lentinus crinitus* (L.: Fr.) Fr., *Samuels 3982,* Feb–Mar 1986 (NY, CAY). FRANCE. PYRENÉES-ATLANTIQUES: Vic. Oloron, Agnos, Le Forêt de

Bugunguc, on immersed ascomycete on bark of *Crataegus* sp., *Samuels (78-72) & Candoussau,* 25 Aug 1978 (PDD 38805). JAPAN. HONSHU: Mt. Fuji, 2nd–3rd Station, Shojinguchi, on sooty mold on wood, *Samuels (83-347),* 5 Sep 1983 (PDD 46525, NY). NEW ZEALAND. North Island. NORTHLAND: Hokianga County, Waipoua State Forest, vic. Forest H.Q., on *Trametes* ? *versicolor* (L.: Fr.) Lloyd, *Samuels (83-74), Matsushima & Petersen,* 11 May 1983 (PDD 46276); AUCKLAND: N of Warkworth, Dome Valley, on *Stereum* sp., *Samuels,* 27 Apr 1976 (PDD 35998); Waitemata City, Waitakere Ranges, off Anawhata Rd, 1000 ft, on *Pseudopanax arboreus* (Murr.) Philipson, *Dingley,* 28 May 1955 (PDD 14602, NY); Waitakere Ranges, Waiatarua, Nature Walk, on resupinate basidiomycete on *Cyathea medullaris* (Forst. f.) Swartz, *Samuels (74-12),* 4 Feb 1974 (PDD 32566); Manukau City, Hunua Gorge, on *Stereum illudens* Berk., *Hughes,* 12 Dec 1963 (PDD 23575).

Notes. The present work documents a much wider distribution for *N. oropensoides* than was previously recognized (Samuels, 1976a). *Nectriopsis oropensoides* is fortunately easily brought into culture. The cultures that I have examined from Europe, USA, French Guiana, Venezuela and New Zealand have been uniform, thus confirming the morphological and cultural homogeneity of this truly cosmopolitan species.

Nectriopsis oropensoides is most closely related to the myxomyceticolous species *N. candicans* and *N. exigua. Nectriopsis oropensoides* is characterized by its small, oblong, spinulose ascospores.

39. Nectriopsis exigua (Patouillard) Gams *in* Gams & van Zaayen, Neth. J. Pl. Pathol. **88:** 73. 1982.

≡*Hypomyces exiguus* Patouillard, Bull. Soc. Mycol. France **18:** 180. 1902.

≡*Nectria myxomyceticola* Samuels, Mycologia **65:** 409. 4 May 1973.

Anamorph. *Verticillium rexianum* (Saccardo) Saccardo, Syll. Fung. **4:** 153. 10 Apr 1886.

Mycelium white, floccose, scant or abundant. Perithecia free or immersed in mycelium, nonstromatic, subglobose, 140–216 × 125–200 μm, or globose, 200–370 μm diam., not changing color in 3% KOH, white to pale yellow, with spinulose, cellular hairs, 20–100 μm long × 3–5 μm wide at base and tapering to the tip. Asci cylindrical to clavate, (30–)40–50(–60) × 2.5–6 μm, apex with a ring. Ascospores oblong, (4–)4.7–5.5(–7) × 2–3 μm, spinulose, colorless.

Habitat. On myxomycete sporangia.

Notes. For illustrations and a complete description see Samuels (1973, as *Nectria myxomyceticola*). *Nectriopsis exigua* is the most common myxomyceticolous species of the genus known from the neotropics.

40. Nectriopsis candicans (Plowright) Maire, Ann. Mycol. **9:** 324. 10 Aug 1911.

≡*Hypomyces candicans* Plowright, Grevillea **11:** 50. Dec 1882.

Anamorph. *Acremonium* sp.

Mycelium white, floccose, scant to abundant. Perithecia immersed in white mycelium, broadly pyriform to globose, (125)–170–210(–250) × (105–)140–165(–240) μm, white, lacking hairs. Asci cylindrical, (36–)45–55(–61) × 3–4(–5) μm, apex with a ring. Ascospores oblong to narrowly ellipsoidal, (5–)6–7.5(–8.7) × 2–3 μm, equally 2-celled; a colorless appendage often visible at each end of each ascospore while still in the ascus, ends of discharged ascospores appearing truncate; smooth, colorless

Habitat. On myxomycete sporangia.

Notes. For illustrations, synonymy and a complete description see Samuels [1971, 1973, as *Nectria candicans* (Plowright) Samuels]. This species was previously known only from the northern hemisphere where it is the most frequently encountered of the myxomyceticolous species (Samuels, 1973). It is now known to occur in the southern hemisphere as well [AUSTRALIA. SOUTH AUSTRALIA: Adelaide, National Park, on myxomycete on *Eucalyptus* sp., *Hughes 1334,* 16 Jul 1963 (PDD 23103, NY); NEW ZEALAND. North Island. COROMANDEL: Thames Co., Coromandel Forest Park, Kauaeranga Valley, on *Physarum* sp., *Samuels 73-168,* 15 Aug 1973 (PDD 32675)].

41. Nectriopsis violacea (Fries) Maire, Ann. Mycol. **9:** 323. 10 Aug 1911.

≡*Sphaeria violacea* Fries, Summa Veg. Scand. **2(2):** 441. 1823.

Anamorph. *Acremonium fungicola* (Saccardo) Samuels, Mycologia **65:** 404. 4 May 1973.

Mycelium white, dense. Perithecia globose to broadly pyriform, (116–)240–275(–390) × (150–) 240–260(–310) μm, violet to purple, not changing color in 3% KOH, with an apical fringe of 10–50 μm long × 5 μm wide hyphal hairs. Asci cylindrical, (40–)50–60(–75) × 3–5 μm, apex with a ring. Ascospores ellipsoidal to oblong, (5–)7–8 × 2.5–3 μm, spinulose, colorless.

Habitat. On aethalia of *Fuligo septica* (L.) Wiggers.

Note. For illustrations, synonymy and a complete description see Samuels [1971, 1973, as *Nectria violacea* (Fries) Fries].

42. Nectriopsis sporangiicola (Samuels) Samuels, comb. nov.

≡*Nectria sporangiicola* Samuels, Mycologia **65:** 416. 4 May 1973.

Anamorph. *Gliocladium* sp.

Mycelium white, floccose. Perithecia pyriform, (130–)140–170(–220) × (100–)120–130 (–160) μm, brownish-yellow, not changing color in 3% KOH. Asci clavate, 30–45 × (3.6–)5–6 (–7) μm, apex with a ring. Ascospores ellipsoidal to oblong, (7–)8–9(–10) × 2–3.5 μm, equally 2-celled, spinulose, colorless.

Habitat. On myxomycete sporangia.

Notes. For illustrations and a complete description see Samuels (1973, as *Nectria sporangiicola*). *Nectriopsis sporangiicola* has previously been known only from the type collection (USA. New Jersey) but has since been recollected in New Jersey [SUSSEX COUNTY: Cedar swamp, High Point State Park, on *Fuligo septica* on moss-covered log, *Rogerson 77-173,* 23 Sep 1977 (NY)] and Brazil [AMAZONAS: Serra Aracá, plateau, N side of North mountain, 1250 m; 00°57′N, 63°21′W, cloud forest, on *Fuligo* sp., *Samuels 461,* 17–22 Feb 1984 (INPA, NY)].

43. Nectriopsis hirsuta (Samuels) Samuels, comb. nov.

≡*Nectria hirsuta* Samuels, Mycologia **65:** 408. 4 May 1973.

Anamorph. None known.

Mycelium white, floccose. Perithecia broadly pyriform, 370–400 × 340–500 μm, yellow, not changing color in 3% KOH, hyphae growing outwardly from perithecial papilla as short hairs 5–10 μm long. Asci cylindrical to clavate, 67–95 × 6–10 μm, apex simple. Ascospores ellipsoidal, 9–13 × 4–5 μm, equally 2-celled, spinulose, colorless.

Habitat. On myxomycete sporangia.

Notes. Synnemata of *Stilbella tomentosa* (Schraeder: Fries) Bresadola var. *ovalispora* (A. L. Smith) Rogerson *in* Samuels (1973) are associated with perithecia of *N. hirsuta* in the type and only known collection. A genetic link between these two morphs has not been proven.

For illustrations and a complete description see Samuels (1973, as *Nectria hirsuta*).

44. Nectriopsis microthecia Samuels, sp. nov. (Fig. 14C–E).

Perithecia minuta, 70–100 × 50–80 μm, superficialia, non stromatica, lutea. Perithecii paries 10 μm crassus. Ascosporae ellipsoideae, (6.3–) 7–8.7(–9) × 2–3.2 μm, 1-septatae. Status anamorphosicus est *Acremonium* sp., conidiis (5–)6.5–10.8(–14) × 1.5–2.2(–3) μm, oblongis vel cylindricis, rectis vel curvatis. Colonia in agaro salmonicolor.

Status anamorphicus *Acremonium*.
Holotypus. Venezuela (VEN: *Samuels 1895*).
Anamorph. *Acremonium* sp.

Mycelium not evident. Perithecia gregarious, superficial, seated on a thin crust of compacted, sometimes pseudoparenchymatous hyphae, crust with perithecia easily removed from the substrate, pyriform, 70–100 × 50–80 μm, with an acute papilla, smooth, collapsing by lateral pinching; orange, not changing color in 3% KOH.

Cells at the surface of the perithecial wall with unevenly thickened walls, and distinctly angular lumina ca. 5 μm in greatest dimension; lumina of adjacent cells joined by minute pores. Perithecial wall 7–10 μm wide, comprising a single region of elliptic, 4–8 × 2 μm cells with walls 0.5 μm thick. Ostiolar area formed of narrow hyphal elements that merge with the periphyses within.

Asci broadly cylindrical, 30–35 × (6.3–)6.8–7.6(–8) μm, apex with a minute ring or appearing simple; 8-spored, ascospores biseriate, completely filling each ascus. Ascospores ellipsoidal to rarely oblong, (6.3–)7–8.7(–9) × 2–3.2 μm, equally 2-celled, not constricted at the septum, smooth, colorless.

Characteristics in Culture. Colonies grown three weeks at 20°C diffuse daylight on CMD, PDA, OA 1.5–2 cm diam., flat, slimy, no aerial mycelium, pale salmon except intensely orange-salmon in center of PDA colonies. Conidiophores arising directly from hyphae on surface

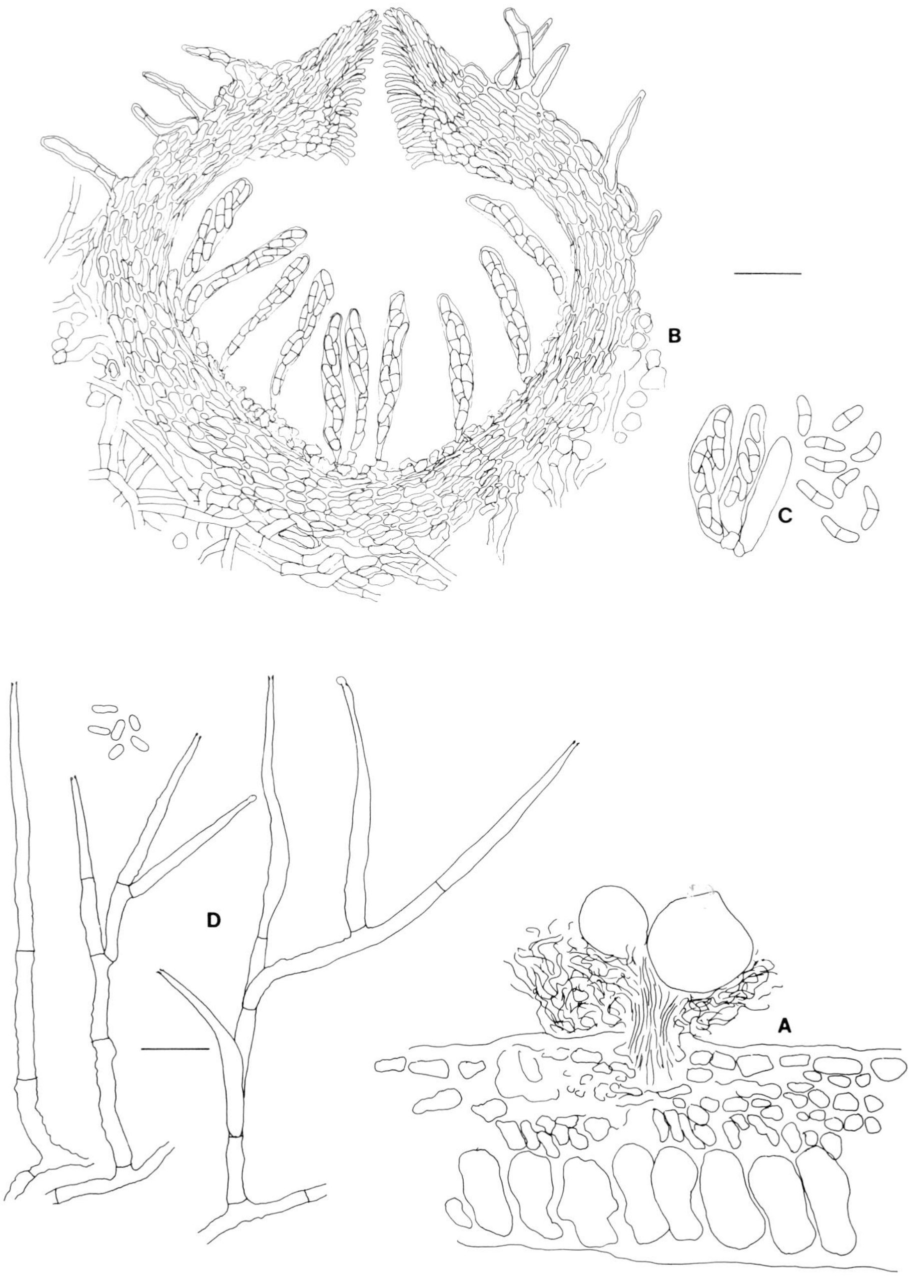

FIG. 16. *Nectriopsis nanocarpa.* **A.** Habit sketch of two perithecia on leaf surface. **B.** Section through a mature perithecium. **C.** Asci and ascospores. **D.** Conidiophores. All drawn from the holotype. Lines = 10 μm.

of colony, micronematous and conidia formed on short, phialidic openings of hyphae; or macronematous, 5–20 μm long, monophialidic and with a single, basal septum or 2-septate and often with a single terminal phialide subtended by a very short phialidic spur. Conidia (5–)6.5–10.8 (–14) × 1.5–2.2(–3) μm, oblong to nearly cylindrical, straight or curved, lacking an obvious basal abscission scar, unicellular, colorless, produced in basipetal succession and held in salmon-colored slime.

Habitat. On bamboo culm, possibly growing on black mycelium.

Known Distribution. Venezuela, known only from the type.

Etymology of the Specific Epithet. Refers to the small size of the perithecia.

Holotype. Venezuela. Amazonas: Dpto. Rio Negro, Cerro de la Neblina, along Río Mawarinuma, just outside Cañon Grande, vic. Neblina base camp, 00°50′N, 66°10′W, ca. 140 m, on bamboo, *Samuels 1895*, Apr, May 1984 (VEN; isotype NY).

Notes. The perithecia of *N. microtheca* and the following species, *N. nanocarpa,* are among the smallest I have seen in the Hypocreales. These species resemble somewhat *N. squamulosa* (see above) a species which itself may be growing on the mycelium of other fungi on decaying debris.

45. Nectriopsis nanocarpa Samuels, sp. nov. (Fig. 16).

Perithecia subglobosa, 50–90 μm diam., setosa, flavida. Perithecii paries ca. 10 μm crassus. Asci clavati, 19–25(–26) × 5–6(–6.5) μm, apice simplices. Ascosporae bacillares, (5.5–)6–7 (–7.5) × 1.6–1.8(–2) μm, rectae vel curvatae, hyalinae, laeves.

Status anamorphicus *Acremonium*?
Holotypus. Venezuela (VEN: *Dumont-VE 1156*).
Anamorph. ?*Acremonium* sp.

Mycelium dense, forming sharply delimited, ± rectangular spots 7 mm long × 2 mm wide on adaxial leaf surface, with fascicles of perpendicular hyphae joining the mycelium on the surface of the leaf with the mycelium within the epidermis of the leaf and there hyphae spreading through epidermal cells. Hyphae of surface mycelium 3–3.5 μm wide with somewhat thickened walls, branched, septate with few free ends. *Acremonium* forming at the edge of each spot and from the surface of perithecia; conidiophores 50–90 μm long × 1.5–2.5 μm at base, macronematous, mononematous, with hemispherical accretions on the wall, unbranched and monophialidic or loosely branched, each branch terminating in a single phialide. Phialides 25–35 μm long × 1.5–2.5 μm wide at the base, straight, with hemispherical accretions scattered on the wall, tip 1 μm wide, not visibly thickened and collarette not flared. Conidia oblong, (2.5–)2.7–4(–5.4) × (1–)1.2–1.8 μm, lacking a visible basal abscission scar, unicellular, colorless.

Perithecia gregarious, up to 100 forming in each spot, partially covered by white mycelium, subglobose, 50–90 μm diam., with a minute, conical papilla, smooth, becoming cupulate when dry; yellow, not changing color in 100% lactic acid; perithecial wall with hyphal hairs scattered around the apex, hairs 20–30 × 2–3 μm, septate, unbranched, smooth, walls slightly thickened.

Surface of perithecial wall obscured by hyphae, forming *textura epidermoidea* with unevenly thickened walls, the lumina of adjacent cells joined by fine pores. Perithecial wall ca. 10 μm wide, comprising a single region of small, nondescript cells, perithecial apex not anatomically distinct from the wall below. Ostiolar region of narrow hyphal elements that merge with the periphyses within. Asci clavate, 19–25(–26) × 5–6(–6.5) μm, apex simple; 8-spored, ascospores biseriate, completely filling each ascus. Ascospores bacillar, (5.5–)6–7(–7.5) × 1.6–1.8(–2) μm, straight or curved, equally 2-celled, not constricted at the septum, smooth, colorless.

Habitat. On living leaf of a monocotyledonous plant, probably a plant parasite.

Known Distribution. Venezuela; known only from the type collection.

Etymology of the Specific Epithet. Refers to the small size of the perithecia.

Holotype. Venezuela. Aragua: Parq. Nacional Henry Pittier, Rancho Grande, path between hotel and hotel's water source, on monocot. leaf, *Dumont (VE 1156), Haines & Samuels,* 3 Jul 1971 (VEN; isotype NY).

Notes. *Nectriopsis nanocarpa* differs from all other species of *Nectriopsis* in being almost certainly a plant parasite. Each white spot is joined to the internal epidermal cells by several "pegs" of highly compacted, vertically oriented hyphae. These hyphae terminate in a pseudoparenchymatous mass of cells that completely fills and apparently destroys one or two epidermal cells.

Nectriopsis nanocarpa is most closely related to *N. microthecia* but differs in having abundant white mycelium, smaller perithecia and a probable plant parasitic habit.

46. Nectriopsis septofusidiae Samuels, sp. nov. (Fig. 17).

Perithecia in mycelio albo immersa, globosa, (95–)115–155(–170) μm diam., non papillata,

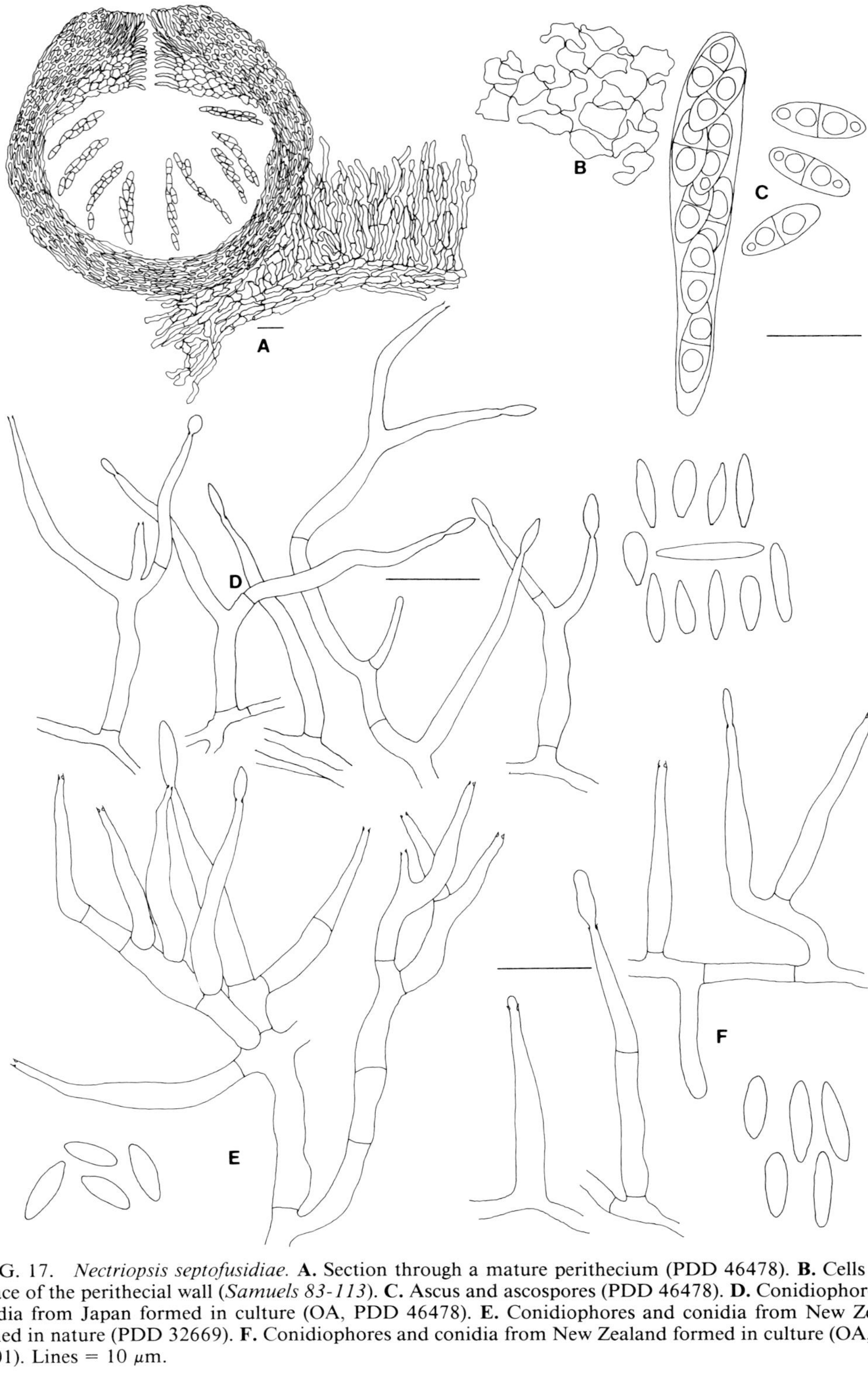

FIG. 17. *Nectriopsis septofusidiae.* **A.** Section through a mature perithecium (PDD 46478). **B.** Cells at the surface of the perithecial wall (*Samuels 83-113*). **C.** Ascus and ascospores (PDD 46478). **D.** Conidiophores and conidia from Japan formed in culture (OA, PDD 46478). **E.** Conidiophores and conidia from New Zealand formed in nature (PDD 32669). **F.** Conidiophores and conidia from New Zealand formed in culture (OA, PDD 46301). Lines = 10 μm.

laevia, flavida. Asci clavati, (33–)38–46(–53) × (7–)7.5–9.5(–10) μm, apice simplices. Ascosporae ellipsoideae, (7–)8.5–10.5(–12) × 2.7–3.5 (–4) μm, laeves, hyalinae.

Status anamorphicus *Septofusidium* cf. *elegantulum* (Pidlop.) W. Gams.
Holotypus. Japan (TNS: *Samuels 83-293*).
Anamorph. *Septofusidium* cf. *elegantulum* (Pidlop.) W. Gams, *Cephalosporium*-artige Schimmelpilze (Hyphomycetes) 147. 1971.

Mycelium white to pale pink, ± radiating from a central point and spreading over the substrate, conidiogenous. Conidiophores 20–30 μm long × 2–3 μm wide at base, colorless, smooth, macronematous, mononematous and unbranched, or irregularly branched and each branch terminating in a single phialide or phialides crowded along the length of the conidiophore branch; phialides 10–20 μm long × 2–3 μm wide at base, sometimes polyblastic with two conidiogenous loci or acropleurogenous with a short phialidic spur subtending the delimiting septum of the terminal phialide. Conidia (6–)5.5–7.2(–9.5) × 2–2.5(–3) μm, ellipsoidal to fusiform to clavate, with or without a minute, flat basal abscission scar, unicellular, colorless, produced in basipetal succession and held in dry chains.

Perithecia gregarious, superficial, seated in mycelium, nonstromatic, globose, (95–)115–155(–170) μm diam., nonpapillate, smooth, collapsing by lateral pinching when dry or not collapsing; yellow, not changing color in 3% KOH. Cells at the surface of the perithecial wall often obscured by hyphae, *textura epidermoidea* with unevenly thickened walls, lumina of adjacent cells joined by minute pores. Perithecial wall ca. 15 μm wide, comprising a single region of indistinct cells; perithecial apex not anatomically distinct from the wall below, ostiolar canal periphysate.

Asci clavate, (33–)38–46(–53) × (7–)7.5–9.5 (–10) μm, apex simple; 8-spored, ascospores partially to completely biseriate, completely filling each ascus. Ascospores ellipsoidal, (7–)8.5–10.5 (–12) × 2.7–3.5(–4) μm, equally 2-celled, not constricted at the septum, smooth, colorless.

Characteristics in Culture. Colonies grown one month at 20°C, 12 hours darkness/12 hours near ultraviolet + cool white fluorescent light on CMD and PDA restricted to the original inoculum; on OA 2–3 mm diam.; aerial mycelium pale pink, with abundant conidial production. Conidiophores and phialides as found in nature.

Habitat. On ascomycetes immersed in woody tissue.

Known Distribution. Japan, New Zealand.

Etymology of the Specific Epithet. Refers to the *Septofusidium* anamorph.

Holotype. Japan. Honshu: Gumma Pref., Katashima-mura, Tone-gun, Jugo Seishi Co. forest, on *Diaporthe* sp., *Samuels 83-293,* 26 Aug 1983 (TNS; isotypes NY, PDD 46478).

Additional Specimens Examined. NEW ZEALAND. AUCKLAND: Waitemata City, Waitakere Ranges, Piha Road, Cowan Track, on immersed pyrenomycete on *Ripogonum scandens* J. R. & G. Forst., *Samuels & Rossman,* 4 Jun 1983 (NY, PDD 46301); Waitakere Ranges, Kitakiki Stream, Marguerite Track, on bark of *Coprosma australis* (A. Rich.) Robinson, *Dingley, Samuels & Haydon,* 30 May 1973 (NY, PDD 31790). TARANAKI: Mt. Egmont National Park, vic. North Egmont Chalet, along track to Waiwhakaiho River, on *Diaporthe* sp. on *Coprosma australis, Samuels,* 1 Oct 1973 (NY, PDD 32669).

Notes. The anamorph of *N. septofusidiae* agrees reasonably well in morphology with the descriptions of *Septofusidium elegantulum* provided by Gams (1971) and Kirk (1983). The growth rate recorded by Gams (1971) for *S. elegantulum* is much faster than I found for the ascospore isolates of two collections (*83-293* from Japan, PDD 46301 from New Zealand) of *N. septofusidiae.* Gams (1971) found conidia with 2–4 cells in cultures of *S. elegantulum*; I have seen only unicellular conidia and Kirk (1983) reported only unicellular conidia for *S. elegantulum.*

Septofusidium elegantulum is known to be hyperparasitic on leaf inhabiting fungi (Gams, 1971; Kirk, 1983). It has not previously been linked to a teleomorph, nor has either of the two additional species included in the genus (Samson, 1974).

Nectriopsis septofusidiae differs from *N. epimyces* in having smaller ascospores.

47. Nectriopsis epimyces Samuels, sp. nov. (Fig. 18).

Nectriopsidi septofusidiae Samuels similis, sed ascosporis (10–)10.5–13(–16.2) × (2–)2.3–3(–3.5) μm.

Status anamorphicus *Septofusidium*?
Holotypus. Indonesia (BO: *Samuels 2191*).
Anamorph. ?*Septofusidium* sp.

Mycelium white, forming spots to 1 mm diam.; hyphae 4 μm wide, straight, branched, smooth, conidiogenous. Conidiophores colorless, smooth,

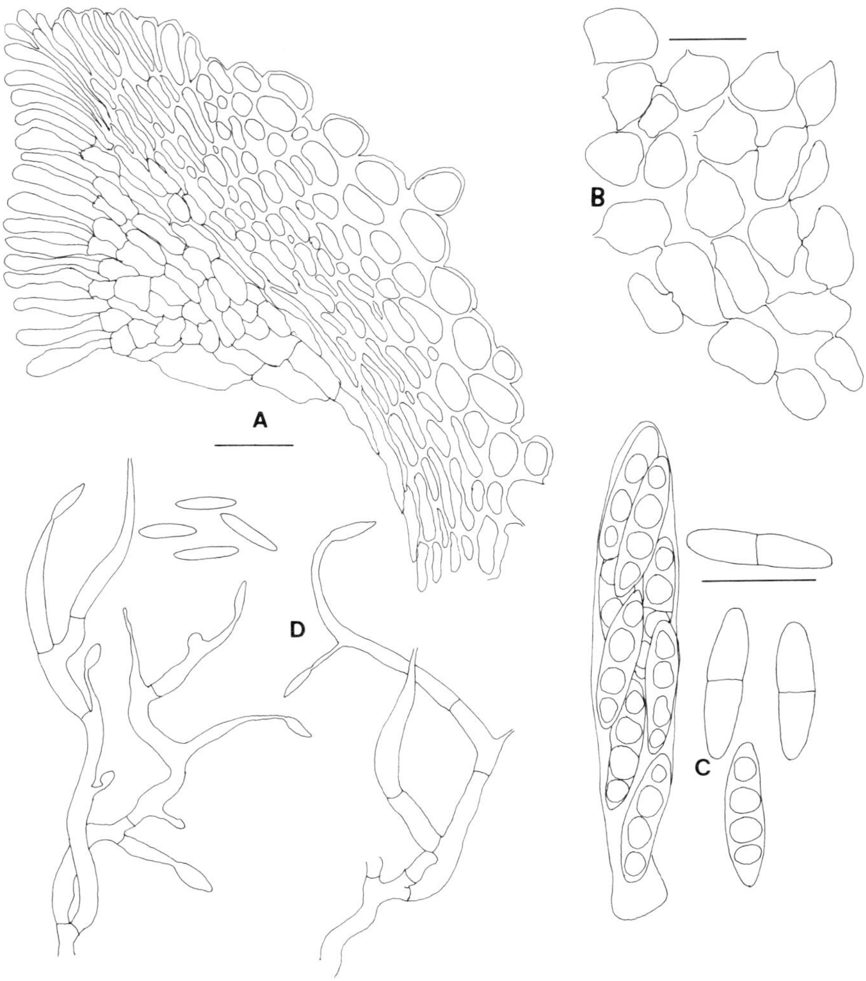

FIG. 18. *Nectriopsis epimyces.* **A.** Part of ostiolar canal and upper portion of wall of a mature perithecium (*Samuels 2050*). **B.** Cells at the surface of the perithecial wall (*Samuels 2473*). **C.** Ascus and ascospores (*Samuels 2473*). **D.** Associated conidiophores and conidia (*Samuels 2225*). Lines = 10 μm.

loosely joined laterally or mononematous, loosely and irregularly branched, terminal branches phialides; phialides smooth, 17–40 μm long × 3 μm wide at base, sharply narrowed below the 1 μm wide tip, polyblastic with 1–2 peg-like, acropleurogenous, phialidic conidiogenous loci forming below the tip; conidiogenous loci lacking obvious periclinal thickening, collarette not flared. Conidia narrowly ellipsoidal, (5.4–)8.2–10.5(–11.7) × (1.8–)2–3(–3.5) μm, with a small, flat, slightly protuberant basal abscission scar, unicellular, hyaline, formed in basipetal succession and held end to end in chains.

Perithecia solitary or in groups of up to 15, partially to completely immersed in white mycelium, stromatic; stroma of densely compacted, 3–4 μm wide, vertically oriented hyphae; perithecia globose, 100–170(–205) μm diam., nonpapillate, smooth, collapsing by lateral pinching; yellow, not changing color in 3% KOH.

Cells at the surface of the perithecial wall with ± angular lumina 7–10 μm in greatest dimension and unevenly thickened walls 1.5–2 μm thick; lumina of adjacent cells joined by narrow pores. Perithecial wall 25–30 μm wide, cells at the exterior elliptic in outline, 6–8 × 4–5 μm, lumina indistinct in shape, walls 1.5–2 μm thick. Perithecial apex formed of narrow hyphal elements; elements becoming progressively narrower toward the ostiolar canal and merging with the periphyses.

Asci clavate, (27–)31–46(–50) × (5.5–)6.3–8.5 μm, apex simple; 8-spored, ascospores biseriate, completely filling each ascus. Ascospores narrowly ellipsoidal, sometimes naviculate, (10–)10.5–13(–16.2) × (2–)2.3–3(–3.5) μm, equally 1-celled, not constricted at the septum, smooth, colorless.

Habitat. On immersed ascomycetes (?Amphisphaeriaceae) on palms, less frequently on immersed ascomycetes on bamboo culms.

Known Distribution. Indonesia (North Sulawesi).

Etymology of the Specific Epithet. Refers to the mycoparasitic habit.

Holotype. Indonesia. North Sulawesi, Dumoga-Bone National Park, vic. 'Hogs Back Camp'; 0°35′N, 123°51′E, 492 m, on *Oxydothis* sp. on arecoid palm, *Samuels 2191,* Oct 1985 (BO; isotype NY).

Additional Specimens Examined. INDONESIA. NORTH SULAWESI: Dumoga-Bone National Park, 'Edwards Camp'; 0°35′N, 123°51′E, 664 m, on immersed ascomycete on arecoid palm, *Samuels 2050,* 6, 8 Oct 1985 (NY); Dumoga-Bone National Park, vic. Project Wallace Base Camp, at confluence of Toraut and Tumpah Rivers; 0°34′N, 123°57′E, 211 m, on ?*Leiosphaerella* on arecoid palm, *Samuels 2473,* Sep, Oct 1985 (BO, NY); same locality, second collection, on bamboo culm, *Samuels 2225* (BO, NY); Dumoga-Bone National Park, vic. camp '1440'; 0°37′N, 123°51′E, 1140 m, on immersed ascomycete on ?arecoid palm, *Samuels 2085,* 3–5 Oct 1985 (NY).

Notes. I was unable to culture this species. The *Septofusidium* anamorph described above was found on three of the five collections cited above. *Nectriopsis epimyces* differs from *N. septofusidiae* in having somewhat larger ascospores.

48. Nectriopsis cupulata (Theissen) Samuels, comb. nov. (Fig. 19).

≡*Byssonectria cupulata* Theissen, Ann. Mycol. **8:** 458. 31 Aug 1910.

Anamorph. None known.

Mycelium white, thin or dense. Perithecia solitary to densely gregarious, with base immersed in mycelium, nonstromatic, globose to broadly pyriform, 130–220 μm diam., nonpapillate, smooth, becoming cupulate or not collapsing when dry; pale yellow, not changing color in 3% KOH.

Cells at surface of perithecial wall obscured by 2–3 μm wide hyphae with walls <0.5 μm thick. Perithecial wall 10–15 μm wide, comprising a single region of flattened to elliptic, 5–10 μm long cells. Perithecial apex formed of a palisade of narrow hyphal elements with ends round, 2–3 μm wide; elements becoming thinner toward the ostiolar canal and merging with the periphyses.

Asci clavate, (32–)36–45(–50) × (4.5–)7–8(–9) μm, apex broad, simple; 8-spored, ascospores uniseriate with overlapping ends or partially to completely biseriate and often arranged transversely within the ascus, completely filling each ascus. Ascospores ellipsoidal, (5.5–)6.3–8.3 (–10) × 2.7–3.5 μm, equally 2-celled, not constricted at the septum, finely spinulose (type) or smooth, colorless.

Habitat. On perithecia and stromata of sphaeriaceous and hypocreaceous pyrenomycetes.

Known Distribution. Brazil.

Holotype. Brazil. Rio Grande do Sul: São Leopoldo, on perithecia of *Nectria gracilipes* (Tulasne & Tulasne) Wollenweber, *Rick,* 1905 (FH!).

Additional Specimen Examined. BRAZIL. AMAZONAS: 0–3 km S of central portion of Serra Araca and 8 km E of Rio Jauari, 60 m; 0°49′N, 63°19′W, on pyrenomycete on recently dead tree, *Samuels 654,* Mar 1984 (INPA, NY).

Notes. *Nectriopsis cupulata* was originally described as forming a compact hyphal stroma within which perithecia were produced at successively higher levels. In the type collection, the stromal aggregates are actually stromata of *Nectria gracilipes* with broken, black synnemata typical of that species. *Nectria gracilipes* itself may be growing over the papilla of a valsaceous fungus and perithecia of *Nectriopsis cupulata* grow on the stroma of *N. gracilipes.* The conidia described by Theissen were not seen.

Nectriopsis cupulata is most closely related to *N. perpusilla*; ascospores of *N. perpusilla* are narrower.

49. Nectriopsis bactridioides (Berkeley & Broome) Samuels, comb. nov. (Fig. 20).

≡*Nectria bactridioides* Berkeley & Broome, J. Linn. Soc. Bot. **14:** 115. 3 Dec 1873.
=*Nectria erinacea* Starbäck, Bih. K. Svenska Vet.-Akad. Handl. **25(III: 1):** 26. 11 Jan 1899.

Anamorph. None known.

Mycelium not evident. Perithecia cespitose in

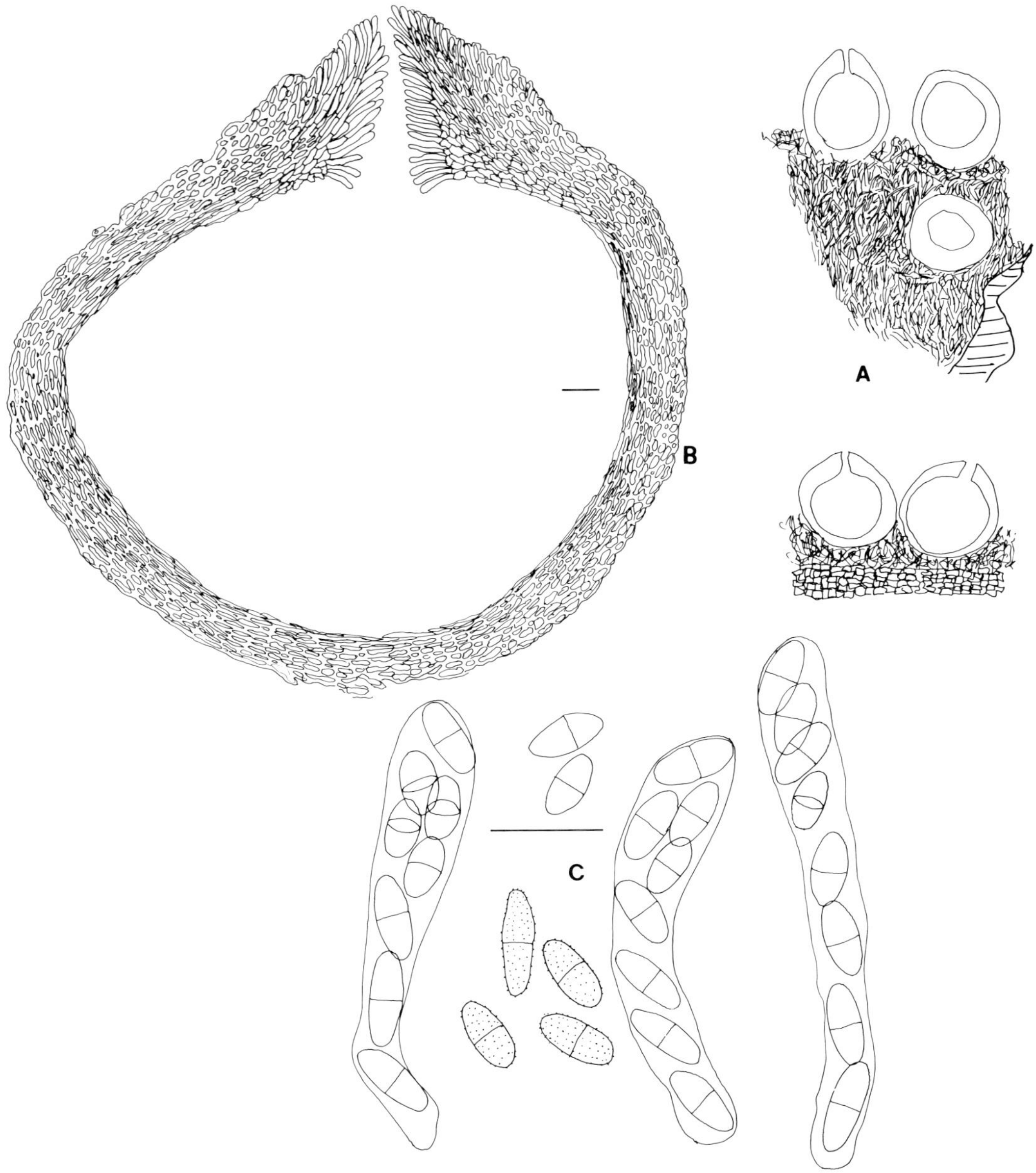

FIG. 19. *Nectriopsis cupulata.* **A.** Two habit sketches of perithecia, the upper sketch with an immersed host perithecium. **B.** Section through a mature perithecium. **C.** Asci and ascospores. All drawn from the holotype. Lines = 10 μm.

groups of several, forming tuberculate to discoidal aggregates, superficial, nonstromatic but difficult to remove from the substrate, pyriform, (110–)123–150(–165) × (80–)90–120 μm, with an acute to subacute apex, smooth, collapsing by lateral pinching when dry; pale orange to yellow, color not changing in 3% KOH.

Cells at surface of perithecial wall lacking a definite outline. Perithecial wall 10–15 μm wide, comprising a single region of small, indistinct cells; perithecial apex not anatomically distinct from the wall below, ostiolar canal periphysate.

Asci clavate, (48–)54–73(–84) × (7–)7.8–9.5 (–11) μm, apex with an obscure ring; 8-spored,

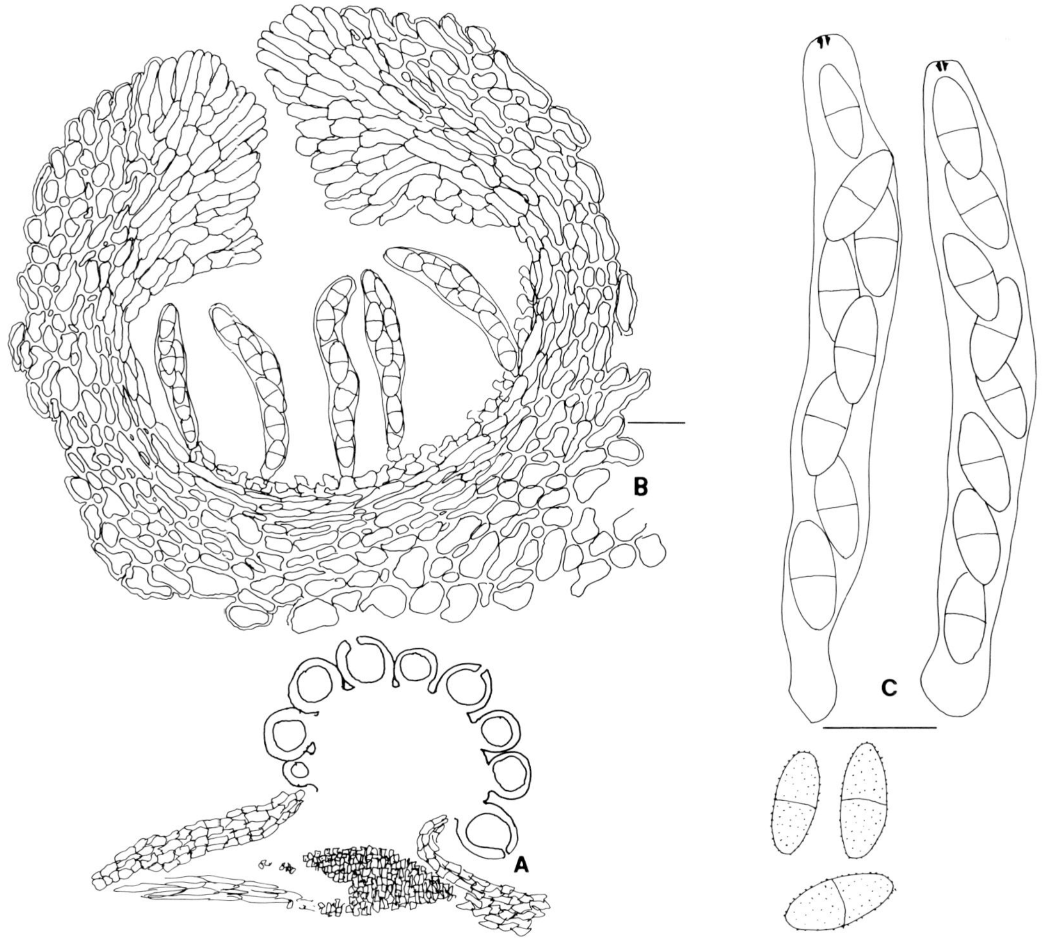

FIG. 20. *Nectriopsis bactridioides.* **A.** Habit sketch of perithecia clustered on the stroma of *Nectria pityrodes.* **B.** Section of a mature perithecium. **C.** Asci and ascospores. All drawn from the holotype. Lines = 10 μm.

ascospores biseriate, completely filling each ascus. Ascospores ellipsoidal, (7–)9–10.5(–11.7) × (4–)4.3–5.5(–6) μm, equally 2-celled, finely spinulose, pale brown to pale yellow.

Habitat. On perithecia and *Myrothecium* sporodochia of *Nectria pityrodes,* and on perithecia of *Nectria rugulosa* Pat. & Gaill.

Known Distribution. Brazil, Japan, Sri Lanka, Trinidad, Venezuela.

Holotype. Ceylon (Sri Lanka). Locality, collector, and date unknown, *no. 58* (K!).

Additional Specimens Examined. VENEZUELA. AMAZONAS: Dpto. Río Negro, Cerro de la Neblina, along Río Mawarinuma, vic. Neblina base camp, ca. 140 m; 00°50′N, 60°10′W; on *Nectria rugulosa, Samuels 1473,* Apr, May 1984 (NY, VEN). TRINIDAD. East of Sangre Grande, on sporodochia, *Seaver,* 10 Apr 1921 (NY). BRAZIL. RIO GRANDE DO SUL: In *Nectria albicante* Starb. parasit., *Malm 336*,* 7/4 93 (S: holotype of *Nectria erinacea*). JAPAN. Bonin Islands, Chichijima, Nakayama-Minamizaki, on *Nectria pityrodes, Doi (D 3552),* 16 Nov 1976 (TNS F 230198); Bonin Islands, Hahajima, Kitamura, on *Nectria pityrodes, Doi* (*D 2743a&b*), 16 Nov 1976 (TNS F 227907); Kitamura, on *Nectria pityrodes, Doi* (*D 2669a & b*), 17 Nov 1976 (TNS F 227834); Kitamura, on *Nectria pityrodes, Doi* (*D 2754*), 17 Nov 1976 (TNS F 227918).

Notes. Tulloch (1972) attributed an unnamed *Myrothecium* anamorph to *Nectriopsis bactridioides* (as *Nectria bactridioides*) following her study of herbarium specimens. The type collection of *Nectriopsis bactridioides* comprises several corticolous, erumpent, hemispheric to tuberculate stromatic aggregates. Perithecia completely obscure the surface of most stromata but two stromata remain free of perithecia and each of these has a plane hymenium of green conidia. These two sporodochia are typical of the *Myrothecium* anamorph of *Nectria pityrodes* and I have concluded that *Nectriopsis bactridioides* is in reality a parasite of that species.

Starbäck (1899) described *Nectria erinacea* as a parasite of *Nectria albicans* Starb. (=*Nectria pityrodes*) and perithecia of the type collection of *N. erinacea* grow upon and completely obscure stromata and perithecia of *Nectria pityrodes.* In the Japanese collections cited above sporodochia and perithecia of *N. pityrodes* are partially to completely covered by perithecia of *Nectriopsis bactridioides.*

Doi (1978) described a *Myrothecium* anamorph (as *Stachylidium*-like) that formed in colonies reportedly derived from ascospores of a collection (TNS: *D 2743a*) of *N. bactridioides* [as *Nectriopsis infusaria* (Cooke & Harkness) Doi]. The anamorph in the dry culture preserved with that collection is typical of *Nectria pityrodes* and I suspect that in his isolation Dr. Doi confused germinating conidia of *N. pityrodes* for ascospores of *N. bactridioides.*

Doi (1978) identified the Japanese collections cited above as *Nectriopsis infusaria* (=*Nectriopsis lasioderma,* see above). *Nectriopsis lasioderma* is distinguished from *N. bactridioides* in having larger and smooth ascospores and also in occurring on sporodochia of a *Fusarium* species.

Perithecia in the collection *Samuels 1473* from Venezuela were seated directly on old perithecia of *Nectria rugulosa,* a species unrelated to *N. pityrodes.* In general this collection agrees with *Nectriopsis bactridioides* but has larger perithecia (155–187 × 110–155 μm), and narrower (9–11.7 × 3.5–4.5 μm), smooth and colorless ascospores. Perithecia are associated with minute synnemata that also arise from old perithecia of *Nectria rugulosa.* Because of these differences I refer this collection to *Nectriopsis bactridioides* with some doubt. Perithecia of the Seaver collection from Trinidad grow on light orange sporodochia but the conidia are colorless and minute, 2 μm or less in diameter; although undoubtedly hypocreaceous, this host is not *Nectria pityrodes.*

50. Nectriopsis byssotecta (Rehm) Samuels, comb. nov. (Fig. 21).

≡*Nectria byssotecta* Rehm, Ann. Mycol. **7:** 539. 31 Dec 1909.

Anamorph. *Gliocladium*-like.

Mycelium conspicuous, white, cottony, spreading somewhat away from host perithecia; hyphae 2–3 μm wide, smooth, septate, much branched with many free ends, conidiogenous. Conidiophores loosely aggregated, 70–80 μm long, ± penicillately branched at the tips; phialides solitary or in 2's or 3's, 10–15 μm long × 3–3.5 μm wide; ± cylindrical, tip with periclinal thickening, not flared. Conidia oblong and straight or slightly curved, to ellipsoidal, 6.5–8(–10) × 2.5–3.5 μm; lacking an obvious basal abscission scar, unicellular, colorless, produced in basipetal succession and held in pale yellow masses.

Perithecia gregarious in groups of 20 or more, superficial, nonstromatic but partially to completely covered by the white mycelium, globose, (130–)160–260(–290) μm diam., nonpapillate, smooth, not collapsing when dry; pale yellow, not changing color in 3% KOH.

Cells at surface at perithecial wall obscured by 4–5 μm wide hyphae. Perithecial wall ca. 10 μm wide, comprising a single region of elliptic to elongated, 10–15 × 3–5 μm cells. Perithecial apex not anatomically distinct from the wall below, ostiolar canal periphysate.

Asci broadly cyindrical to clavate, (38–)45–65(–85) × (5.5–)7.5–11.5(–13) μm, apex broad, simple; 4–8-spored, ascospores irregularly arranged within the ascus, completely filling each ascus. Ascospores ellipsoidal, (7–)9.5–13(–14) × (3–)3.5–4(–5) μm, equally 2-celled, not constricted at the septum, smooth, colorless.

Habitat. On ascomycetes, often on perithecia of *Nectria anacardii* Rehm and *N. jungneri* Henn.

Known Distribution. Brazil, French Guiana, Indonesia, Venezuela.

Holotype. Brazil. Para: An Rinde von *Anacardium occidentale* L., *C. F. Baker,* 24 Feb 1909 (S!; isotype FH: herb. Höhnel!).

Additional Specimens Examined. VENEZUELA. AMAZONAS: Dpto. Río Negro, Cerro de la Neblina, along Río Mawarinuma, just outside Cañon Grande, vic. Neblina base camp; 00°50′N, 66°10′W, 140 m, on *Nectria jungneri, Samuels 1148,* Apr–May 1984 (NY, VEN); same collecting data, second collection *Samuels 1652* (NY, VEN); same locality, on immersed pyrenomycete on live vine, *Samuels 1711,* Apr–May 1984 (NY, VEN). FRENCH GUIANA. Mt. Galbao, ca. 15 km SW of Saül, 600–650 m, on *Nectria jungneri, Samuels (3107) & Boise,* 19 Jan 1986 (CAY, NY). IN-

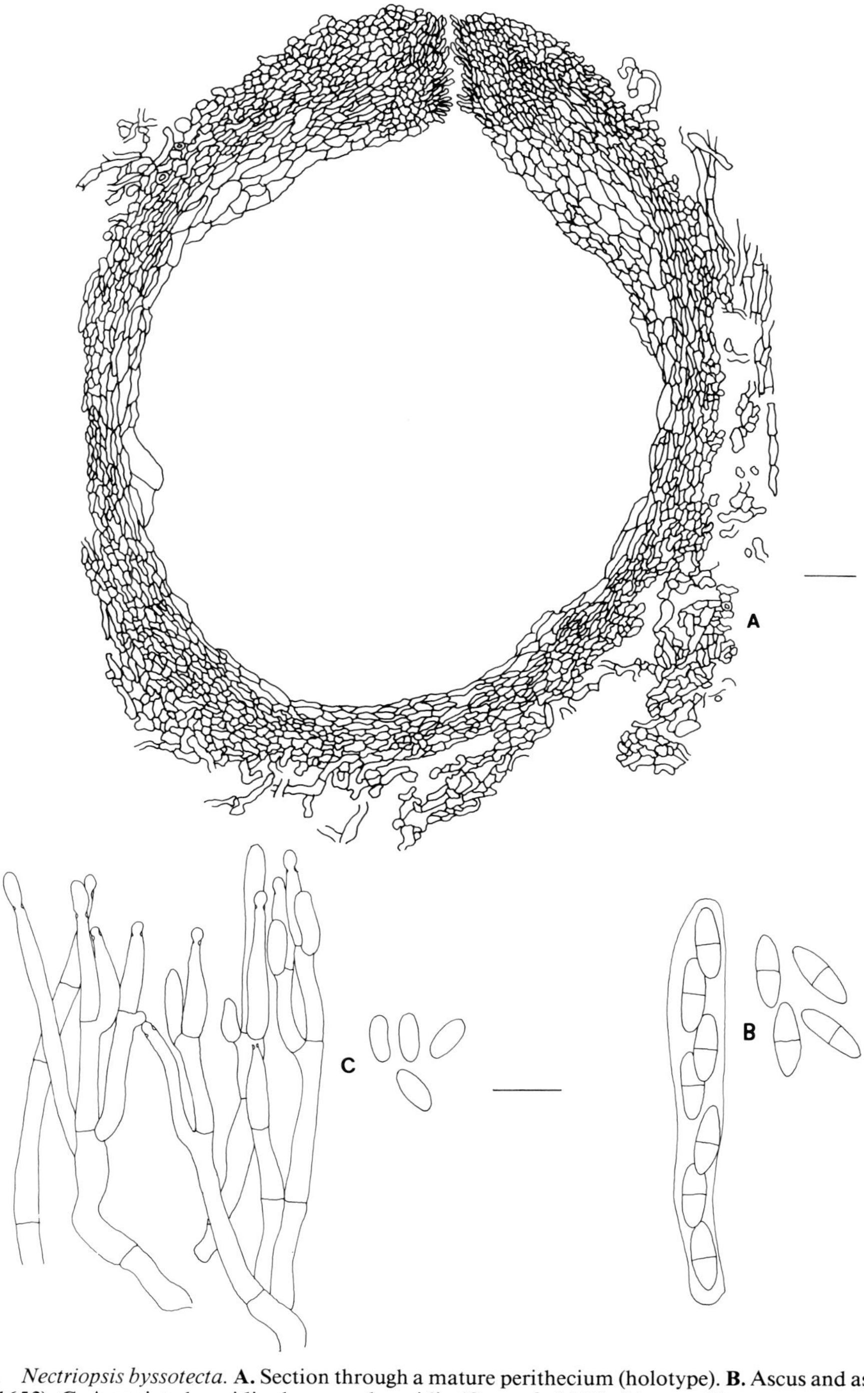

FIG. 21. *Nectriopsis byssotecta.* **A.** Section through a mature perithecium (holotype). **B.** Ascus and ascospores (*Samuels 1652*). **C.** Associated conidiophores and conidia (*Samuels 1652*). Lines = 10 μm.

DONESIA. NORTH SULAWESI: Eastern Dumoga-Bone National Park, between 'Maddison Camp' (00°36'N, 123°50'E) and '1440 Camp' (00°37'N, 123°51'E), 980–1100 m, on *Nectria jungneri, Samuels 2151,* 5 Oct 1985 (BO, NY).

Notes. *Nectriopsis byssotecta* has not been cultured. The same anamorph was found on four collections of *N. byssotecta*: the holotype and *Samuels 1652* (Venezuela), *Samuels 3107* (French Guiana), and *Samuels 2151* (Indonesia).

Nectriopsis byssotecta is morphologically similar to *N. albofulta* but differs in that ascospores of *N. albofulta* are striate, and there are scattered, enlarged cells or circular hairs at the apex of perithecia of that species.

51. Nectriopsis oxyspora Samuels, sp. nov. (Fig. 22A–C).

Perithecia pyriformia, 140–175 × 110–155 μm, glabra, flavida. Perithecii paries 10–15 μm crassus. Asci clavati, 57–65 × 8–10 μm, apice simplices. Ascosporae fusiformes, (13.5–)14.8–17.3 (–18) × 3.5–4.2(–4.5) μm, utrinque acutae, laeves, hyalines.

Status anamorphicus *Septofusidium*-similis.
Holotypus. Brazil (INPA: *Samuels 152*).
Anamorph. *Septofusidium*-like.

Mycelium white, dense, spreading slightly away from the host, with many erect hyphae and conidiophores, hyphae 3–4 μm wide, septate, branched, smooth or spinulose, conidiogenous. Conidiophores arising directly from hyphae, macronematous, mononematous, 100–150 × 3–3.5 μm; spinulose, straight, wall perceptibly thickened, colorless, unbranched or sparingly branched, each branch bearing a terminal phialide and frequently with 1–2 levels of acropleurogenous phialides. Terminal phialide 16–33 × 2.5–3.5 μm, cylindrical, straight, spinulose, wall thickened, slight periclinal thickening visible at the tip, collarette not flared. Conidia not seen with certainty.

Perithecia densely gregarious, with base immersed in white mycelium, nonstromatic, easily removed from the substrate, pyriform, 140–175 × 110–155 μm; with a 50–70 μm wide, blunt apex; smooth, collapsing by lateral pinching when dry; yellow, not changing color in 3% KOH.

Surface of perithecial wall completely obscured by hyphae. Perithecial wall 10–15 μm wide, comprising a single region of nondescript, small cells with walls 0.5 μm thick. Perithecial papilla of narrow, 2 μm wide hyphal elements, hyphal elements merging with periphyses at the interior and growing outwardly 3–5 μm at the surface of the papilla.

Asci clavate, 57–65 × 8–10 μm, apex simple; 8-spored, ascospores biseriate, completely filling each ascus. Ascospore fusiform, (13.5–)14.8–17.3(–18) × 3.5–4.2(–4.5) μm, ends acute, equally 2-celled, not constricted at the septum, colorless, smooth.

Habitat. On old stroma of ?Hypocreales.

Known Distribution. Brazil, known only from type collection.

Etymology of the Specific Epithet. Refers to the ascospores that are acute at each end.

Holotype. Brazil. Amazônas: Pico Rondon, upper vine forest immediately below summit, on ?*Nectria* on bamboo culm, *Samuels 152,* 4 Feb 1984 (INPA, NY).

Notes. The relatively large, fusiform ascospores and the presumed anamorph characterize this species.

Although abundant conidiophores accompany perithecia of the type collection of *N. oxyspora*. I did not find conidia adherent to the conidiogenous cells. The few conidia found in the mycelium around the conidiophores were ellipsoidal to oblong, 1–3-septate, colorless and smooth but they were not necessarily produced from the conidiophores described above. I have referred these conidiophores to *Septofusidium* simply because of the presence of acropleurogenous phialides on colorless, monophialidic *Acremonium* conidiophores.

52. Nectriopsis albofulta (Petch) Samuels, comb. nov. (Fig. 22D, E).

≡*Nectria albofulta* Petch, Ann. Roy. Bot. Garden, Peradeniya **7:** 128. May 1920.

Anamorph. None known.

Mycelium white, dense, filling fissures in bark. Perithecia seated on or up to half immersed in the mycelium, globose, 150–250 μm diam., with a minute, acute papilla, smooth, collapsing by lateral pinching when dry; yellow, becoming pale roseous in 3% KOH; with aculeolate, 0–2-septate, thin-walled, spinulose, unbranched hairs or modified cells 20–30 μm long × 10 μm wide basally arising from around apex of immature perithecia, not noted on mature perithecia.

FIG. 22. A–C. *Nectriopsis oxyspora.* **A.** Section through a mature perithecium. **B.** Ascus and ascospores. **C.** Conidiophores and conidia associated with perithecia. A, C drawn from the holotype. D, E. *Nectriopsis albofulta.* **D.** Section through the apex of a mature perithecium. **E.** Ascospores. D, E drawn from the holotype. Lines = 10 μm.

Cells at surface of perithecial wall nearly circular in outline, 10–15 μm diam., walls ca. 0.5 μm thick. Perithecial wall 10–15 μm wide, comprising two regions; cells of outer region elliptic in section, 10–15 × 5–7 μm, wall <0.5 μm thick; inner region comprising a few layers of flattened and compressed cells. Perithecial apex not anatomically distinct from the lateral wall, ostiolar canal periphysate.

Asci clavate, (33–)36–47(–56) × (5–)6–10(–12) μm, apex simple; 8-spored, ascospores biseriate, completely filling each ascus. Ascospores ellipsoidal, (6.3–)8.4–10.5(–11) × 3.5–4(–4.2) μm, equally 2-celled or septum supramedian and basal cell slightly attenuated, not or slightly constricted at the septum, striate, colorless.

Habitat. On bark of *Beloperone oblongata* Lindl., possibly on *Fusarium* sp. and/or *Nectria gracilipes.*

Known Distribution. Sri Lanka, People's Republic of China.

Holotype. Ceylon (Sri Lanka). Central: Peradeniya, on *Beloperone oblongata, Petch 4958,* January 1917 (K!).

Additional Specimen Examined. CHINA (PEOPLE'S REPUBLIC). GUANGDONG: Hainan, Ting-an, on *Nectria gracilipes, S. Q. Deng 7561,* 8 Dec 1934 (BPI, HMAS, NY).

Notes. The Chinese specimen of *N. albofulta* was found on three isotypes of *Chromocreopsis epinectria* (≡*Nectriopsis epinectria*). The two species are easily distinguished as perithecia of *N. epinectria* (see below) are enveloped by ferrugineous mycelium and contain brown and coarsely striate ascospores.

53. Nectriopsis puiggarii (Spegazzini) Samuels, comb. nov. (Fig. 23A–D).

≡*Nectria puiggarii* Spegazzini, Bol. Acad. Nac. Cienc. Córdoba **11:** 526. Jul 1889.

≡*Cucurbitaria puiggarii* (Spegazzini) O. Kuntze, Rev. Gen. Pl. **3(2):** 461. 28 Sep 1898.

=*Nectria sulcispora* Petch, Ann. Roy. Rot. Gard. Peradeniya **7(2):** 129. 1920.

=*Nectria striatula* Yates, Philipp. J. Sci. **12:** 373. Nov 1917.

Anamorph. *Acremonium* sp.

Mycelium white, cottony, spreading over surface of host stroma, conidiogenous. Conidiophores 20–35 μm long × 3–4 μm wide at base, macronematous, mononematous, colorless, smooth, unbranched or irregularly branched, each branch comprising a single phialide. Phialides 20–25 μm long × 1 μm wide at tip × 1.5–2 μm wide at base, straight, smooth, monoblastic, periclinal thickening of tip at most barely visible, collarette not flared. Conidia oblong, 3–5 × 1–1.5 μm, lacking an obvious basal abscission scar, unicellular, colorless, produced in basipetal succession and held in a colorless drop of liquid at the tip of each phialide.

Perithecia densely gregarious, forming within white mycelium with upper ¼ of each perithecium free, pyriform, 220–280 μm high × 185–250 μm wide, with an obtuse apex, smooth, collapsing by lateral pinching or not collapsing when dry; amber-colored, not changing color in 3% KOH.

Cells at the surface of the perithecial wall forming *textura epidermoidea,* walls of cells <0.5 μm thick. Perithecial wall 20–30 μm wide, comprising a single region of ± elliptical to flattened, 5–7 μm long × 2–3 μm wide cells with walls ≤0.5 μm thick. Perithecial apex not anatomically distinct from the lateral wall, cells at the surface of the papilla ± enlarged, ca. 5 μm wide, ostiolar canal periphysate.

Asci broadly cylindrical to narrowly clavate, (40–)50–75(–80) × (5–)7.5–12.9(16) μm, apex simple; 6–8-spored, ascospores uniseriate with overlapping ends to biseriate, completely filling each ascus. Ascospores ellipsoidal to fusiform, (10.8–)11.3–14.2(–16) × (4–)4.5–6.2(–7.5) μm, equally 2-celled, slightly constricted or not constricted, with coarse, ridge-like striations, colorless.

Characteristics in Culture. Colonies grown 10 days at 20°C, 12 hours dark/12 hours near ultraviolet + cool white fluorescent light. CMD, PDA: 3 cm diam.; OA: 5.5 cm diam. Colonies on all media flat, colonies on CMD translucent, felty; colonies on PDA and OA opaque, densely felty. Conidial production abundant on all media; conidiophores arising from 2.5–3 μm wide aerial hyphae and, on OA, from poorly developed hyphal, rope-like strands. Condiophores macronematous, mononematous, unbranched, 70–115 μm long × 2.5–3.5 μm wide at base, (1–)3–4(–5)-septate, straight, smooth, monophialidic. Phialides 25–60 μm long, straight, smooth, tip slightly thickened or not thickened, not flared. Conidia ellipsoidal, 6–8(–13) × 2.5–3(–3.5) μm, lacking an obvious basal abscission scar, unicellular, arising in basipetal succession

FIG. 23. *Nectriopsis puiggarii.* **A.** Section through a mature perithecium (*Dumont-PE 763*). **B.** Ascus and ascospores (*Dumont-PE 763*). **C.** Conidiophores and conidia from Peru formed in nature (*Dumont-PE 763*). **D.** Conidiophores and conidia from New Zealand formed in culture (PDD 46428). **E.** *Nectriopsis epinectria,* ascus and ascospores, drawn from an isotype (NY). Lines = 10 μm.

and held in a colorless drop of liquid at the tip of each phialide.

Habitat. On stromata of Xylariaceae.

Known Distribution. Brazil, Jamaica, New Zealand, Peru, Philippine Islands, Sri Lanka.

Holotype. Brazil. Apiahy, on *Xylaria hippoglossa* Speg., *Puiggari 2892,* v.1888 (LPS 1108!).

Additional Specimens Examined. JAMAICA. On *Xylaria* sp., *Wright,* 1909 (FH). PERU. HUANUCO: Ca. 38 km from Tingo María, on the Tingo María–Pucallpa Rd., ca. 5200 ft, on *Xylaria* sp., *Dumont (PE 763), Carpenter, Sherwood & Buriticá,* 4 Jul 1976 (NY). BRAZIL. AMAZÔNAS: Plateau of Serra Acaca, N side of north mountain, 00°57′N, 63°21′W, 1250 m; cloud forest, on *Kretzschmaria* sp., *Samuels 333, Prance & Pipoly,* 17–22 Feb 1984 (INPA, NY); same collecting data, second collection, on *Hypoxylon* (*Samuels 340*), *Samuels 341* (INPA, NY). SRI LANKA (Ceylon). CENTRAL: Hakgala, on decaying *Xylaria,* collector unknown (?*Petch*), *nr. 4577a,* May 1913 (K, holotype of *N. sulcispora*). PHILIPPINE ISLANDS. Alabat Island, host not given, *E. D. Merrill 10519,* 21–30 Dec 1916 (K, isotype of *Nectria striatula*). NEW ZEALAND. AUCKLAND: Waitemata City, Waitakere Ranges, Fairy Falls Track off Mountain Rd., on *Hypoxylon deustum* (Hoffm.) Grev., *Samuels & Petersen,* 29 Apr 1983 (PDD 46248).

Notes. Colony characteristics reported above are derived from the New Zealand collection, the only collection that I have been able to culture. The naturally formed anamorph is described from the Peruvian collection.

54. Nectriopsis epinectria (Teng) Samuels, comb. nov. (Fig. 23E).

≡*Chromocreopsis epinectria* Teng, Sinensia **6:** 189. Apr 1935.

Anamorph. None known.

Mycelium ferrugineous around the perithecia and white elsewhere, dense, spreading somewhat, apparently overgrowing host perithecia; hyphae branched, septate, smooth, 3–5 μm wide. Perithecia cespitose in tuberculate aggregations, completely immersed within ferrugineous mycelium, nonstromatic, in groups of 10–25, globose, ca. 175 μm diam., nonpapillate, smooth, collapsing by lateral pinching, orange, not changing color in 3% KOH. Cells at surface of perithecial wall completely obscured by hyphae; hyphae branched, septate, smooth, 3–5 μm wide, with few free ends. Perithecial wall ca. 15 μm wide, comprising a single region of small, nondescript cells. Perithecial apex not anatomically distinct from the lateral wall, ostiolar canal periphysate.

Asci cylindrical, (37–)41–50(–52) × 4–5.5(–6.3) μm, apex simple; 8-spored, ascospores uniseriate, sometimes transversely oriented, completely filling each ascus. Ascospores oblong, 5.5–6.5 (–7.2) × (3–)3.5–3.8(–4) μm, equally 2-celled, slightly constricted, translucent brown, coarsely striate.

Habitat. On perithecia of *Nectria gracilipes.*

Known Distribution. People's Republic of China, known only from the type.

Holotype. China (People's Republic of China). Guangdong: Hainan, Ting-an, on *Nectria, S. Q. Deng 7361,* 8 Dec 1934 (BPI!, NY!).

Notes. The original collection of *Chromocreopsis epinectria* has two species of *Nectriopsis* parasitic on *Nectria gracilipes. Nectriopsis epinectria* is distinctive because of its ferrugineous mycelium and brown, coarsely striate ascospores; the original description was unequivocal, leaving no doubt as to the identity of the species. The second hyperparasite is *N. albofulta* (see above). The original collection of *Chromocreopsis epinectria* was divided at least three times, with a portion remaining in China and other parts distributed to BPI, MO, and NY. The mycological collections of MO are now deposited at BPI. There are now collections with data of the holotype deposited in BPI, HMAS (Beijing), and NY. The portions in BPI and NY have both *N. epinectria* and *N. albofulta.* The portion in HMAS has only *N. albofulta.*

For additional illustrations of *N. epinectria* see Teng (1935, fig. 2a, b).

55. Nectriopsis epimycota Samuels, sp. nov. (Fig. 24).

Perithecia globosa, 100–160 μm diam., non papillata, laevia, flavida. Asci clavati, (25–)28–38(–45) × (4–)4.5–5.5(–6.5) μm, apice simplices.

→

FIG. 24. *Nectriopsis epimycota.* **A.** Cells at the surface of the perithecial wall (holotype). **B.** Ascus and ascospores (holotype). **C.** Conidiophores and conidia formed in nature (holotype). **D.** Conidiophores and conidia formed in culture (holotype). **E.** Cells at the surface of the perithecial wall (*Fife 4050*). **F.** Ascus and ascospores

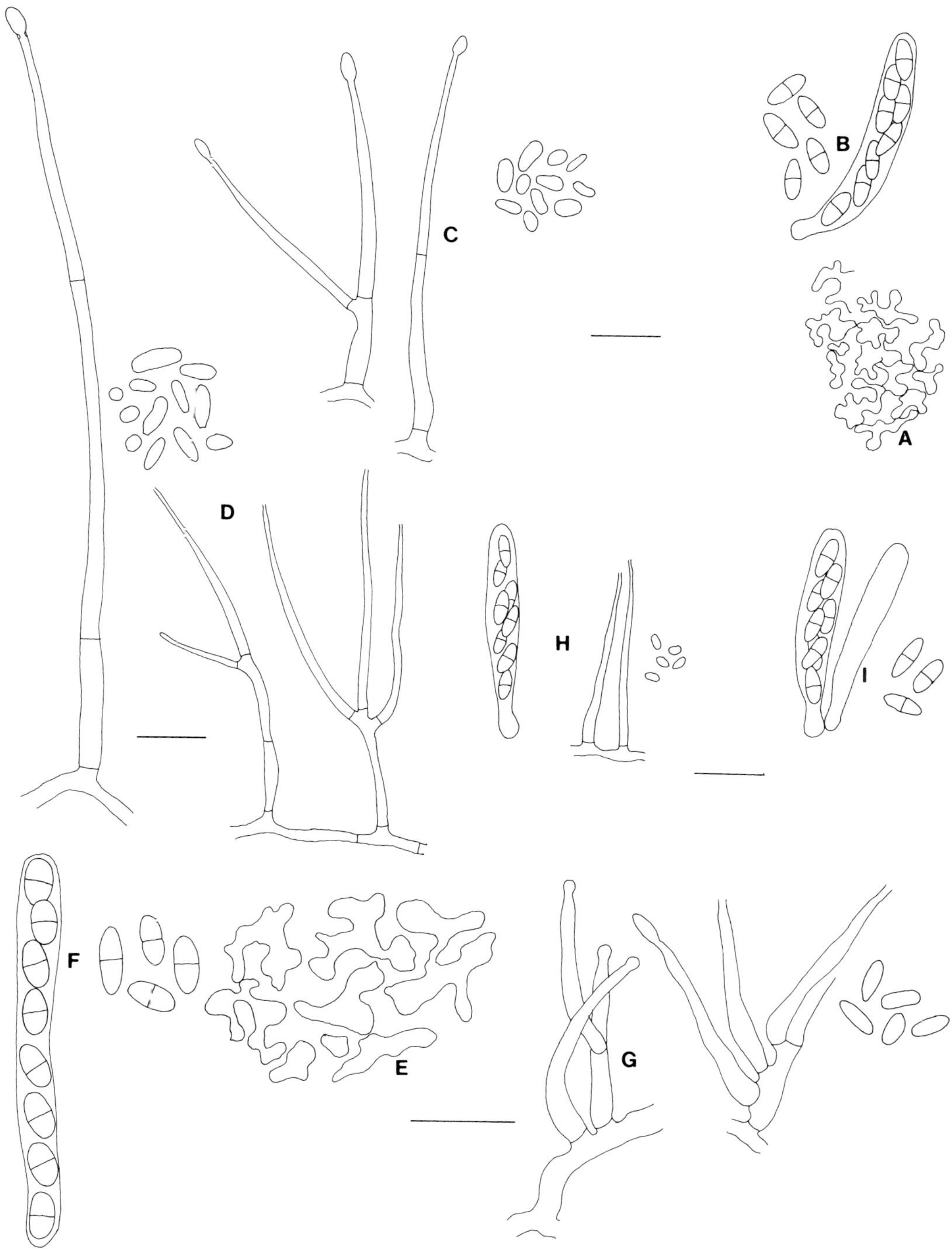

(*Fife 4050*). **G.** Conidiophores and conidia formed in nature (*Fife 4050*). **H.** Ascus and ascospores (*Samuels 3651A*). **I.** Conidiophores and conidia formed in nature (*Samuels 3651A*). **J.** Asci and ascospores (*Samuels 4457*). Lines = 10 μm.

Ascosporae ellipsoideae vel oblongae, (4.5–)5.6–6.3(–7.2) × (1.5–)1.8–2.7(–3) μm, laeves, hyalinae. Conidiophorae solitariae erectae simplices, aut eas *Acremonii* simulantes aut apice penicillatae vel verticillatae, 50–120 μm longae. Conidia in agaro culta subglobosa, 2.5–4.2(–4.5) × (1.8–)2–2.7 μm, aut eseptata aut ellipsoideo-oblonga, 2–3-septata, (4.5–)5.5–8(–10) × (2–)2.3–3.3(–3.5) μm.

Status anamorphicus *Acremonium* sp.
Holotypus. French Guiana (NY: *Samuels 3463*).
Anamorph. *Acremonium* sp.

Mycelium not evident, or white, thin and with 2–2.5 μm wide hyphae, spreading and conidiogenous. Conidiophores arising as lateral branches of hyphae, macronematous, mononematous, straight, smooth, 0–1-septate, monophialidic, 15–35 μm long × 2–2.5 μm wide at base, tip 0.5 μm wide at tip, collarette slightly thickened, not flared. Conidia oblong to ellipsoidal, 2.7–4.3(–5) × 1–1.5(–2) μm, lacking an obvious basal abscission scar, unicellular, colorless.

Perithecia gregarious, superficial, nonstromatic, difficult to remove from the substrate, globose, 100–104(–160) μm diam., nonpapillate, smooth, collapsing by lateral pinching; yellow, not changing color in 3% KOH.

Cells at the surface of the perithecial wall *textura epidermoidea* with unevenly thickened walls, lumina of adjacent cells joined by minute pores. Perithecial wall ca. 10 μm wide, comprising a single region of small, indistinct cells. Perithecial apex not anatomically distinct from the wall below; ostiolar canal periphysate.

Asci clavate, (25–)28–38(–45) × (4–)4.5–5.5 (–6.3) μm, apex simple; 8-spored, ascospores uni- to biseriate, completely filling each ascus. Ascospores ellipsoidal to oblong, (4.5–)5–6.3 (–7.2) × (1.5–)1.8–2.7(–3) μm, equally 2-celled, not constricted at the septum, smooth, colorless.

Characteristics in Culture. Colonies grown one week at 20°C, diffused daylight on CMD restricted to the original inoculum, very pale buff, with scant aerial mycelium. Conidiophores arising abundantly from the surface of the colony, as lateral branches of hyphae, macronematous, mononematous, straight, smooth. Two morphologically distinct types of conidiophores and conidia formed. Type one: conidiophores 50–60 μm long, unbranched or with phialides held in a terminal whorl of 2–3; conidia 2.5–4.2(–4.5) × (1.8–)2–2.7 μm, subglobose, lacking an obvious basal abscission scar, unicellular, colorless. Type two: conidiophores 60–120 μm long, unbranched, 2–3-septate; conidia (4.5–)5.5–8(–10) × (2–)2.3–3.3(–3.5) μm, oblong to ellipsoidal with a broad, protuberant, flat, basal abscission scar. All phialides 30–40 μm long × 1.5 μm wide at base, tip ca. 1 μm wide, collarette thickened or not, unflared. Conidia held in colorless drops of liquid.

Habitat. On ascomata and mycelium of various ascomycetous fungi including *Bertia, Kretzschmaria* and *Penzigia.*

Etymology of the Specific Epithet. Refers to the hyperparasitic habit.

Holotype. French Guiana. Vic. Saül, elev. 200 m, on *Kretzschmaria* sp., *Samuels 3463,* Feb 1986 (NY; isotype CAY).

Additional Specimens Examined. FRENCH GUIANA. Vic. Saül, elev. 200 m, on fungus on sheathing base of palm, *Samuels 3651A,* Feb 1986 (NY); vic. Cayenne, on *Penzigia* sp., *Samuels (4457) & Feuillet,* 23 Mar 1986 (CAY, NY). BRAZIL. AMAZONAS: 70 km S of Humaita on rd. to Porto Velho; 0.5–3 km E of hwy., on *Bertia moriformis* (Tode: Fr.) de Not., *Fife (4050), McFarland, Texiera, Nelson, Santos, Mota & Gomes,* 5 May 1982 (NY).

Notes. Characteristics of the anamorph are taken from the holotype, which was grown in pure culture from ascospores. None of the paratypes was grown. The teleomorphs of all of the cited specimens indicated a single species but anamorphs associated with two of them (*Fife 4050,* Fig. 24E–G; *Samuels 3651 A,* Fig. 24H, I) do not seem to be the same species and are different again from the anamorph obtained in culture from the holotype. Conidia were not found on the holotype specimen, nor were they found on the specimen *Samuels 4457.*

56. Nectriopsis flavella (Patouillard) Samuels, comb. nov. (Fig. 25A–C).

≡*Nectria flavella* Patouillard, Bull. Soc. Mycol. France **16:** 186. 1900.

Anamorph. None known.

Mycelium not apparent. Perithecia solitary or cespitose in groups of ≤10, superficial, seated on a basal stroma, globose, 230–360 μm diam., nonpapillate, slightly scaly, not collapsing when dry;

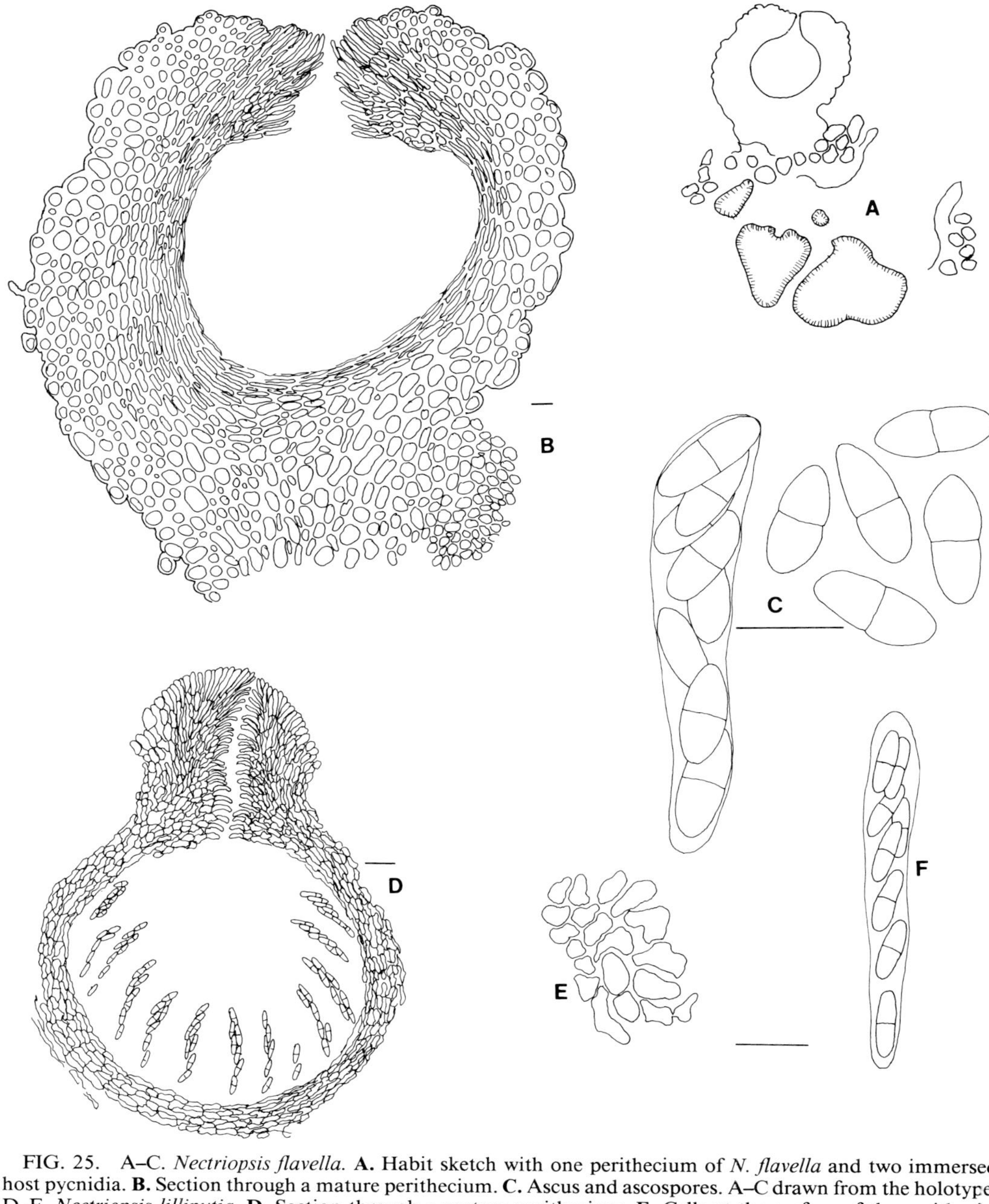

FIG. 25. A–C. *Nectriopsis flavella.* **A.** Habit sketch with one perithecium of *N. flavella* and two immersed host pycnidia. **B.** Section through a mature perithecium. **C.** Ascus and ascospores. A–C drawn from the holotype. D–F. *Nectriopsis lilliputia.* **D.** Section through a mature perithecium. **E.** Cells at the surface of the perithecial wall. **F.** Ascus and ascospores. D–F drawn from the holotype. Lines = 10 μm.

pale orange with lighter colored scales, not changing color in 3% KOH.

Cells at the surface of the perithecial wall circular to elliptical in outline, 10–20 μm diam., walls 2–3 μm thick. Perithecial wall 50–80 μm wide, comprising two regions; outer region 60–70 μm wide, lumina of cells elliptic, 10–20 × 5–10 μm, walls 2–3 μm thick; inner region ca. 15

μm wide, lumina of cells 15–30 × 3 μm, walls 1–2 μm thick. Perithecial apex not anatomically distinct from the wall below, ostiolar opening formed by 3–4 μm wide hyphal elements with round tips, hyphal elements merging with periphyses.

Asci clavate, (39–)41–49(–53) × (6–)8–12(–13) μm, apex broad, simple; 6–8-spored, ascospores multiseriate above, 1-seriate below or multiseriate throughout. Ascospores (10–)11–12.5 (–13) × 4.5–5.5(–6) μm, ellipsoidal, equally 2-celled, not constricted at the septum, smooth, colorless.

Habitat. On herbaceous debris, possibly on fungi.

Known Distribution. Guadeloupe, known only from type collection.

Holotype. Guadeloupe. Grand Etang, on *Guazuma ulmifolia* Lam., *Duss,* date not given (FH!; isotype NY!).

Notes. This is a strange species for which I see no close relationships despite the morphological similarity of the perithecia to perithecia of other species discussed in this paper. Cells of the perithecial wall appeared to be only loosely joined. Perithecia of the holotype and isotype, when placed in a drop of water, were immediately crushed when the cover-slip was put in place.

Although the species is described as growing on a herbaceous substrate, the perithecia are actually growing on perithecia of an immersed, possibly diatrypaceous, ascomycete.

57. **Nectriopsis lilliputia** Samuels, sp. nov. (Fig. 25D–F).

Nectriopsidi septofusidiae Samuels similis sed ascosporis (7.2–)7.9–9.2(–10) × ca. 3 μm, nec non mycelio haud visibili; *N. squamulosae* (Ellis) Samuels similis sed ascosporis brevioribus et angustioribus, nec non asci apice simplici.

Status anamorphicus ignota.
Holotypus. Brazil (INPA: *Samuels 1138*).
Anamorph. None known.

Mycelium thin, white, spreading. Perithecia gregarious, superficial, nonstromatic, easily removed from the substrate, globose, 145–175 × 110–135 μm, nonpapillate, smooth, collapsing by lateral pinching; pale yellow, nearly white, not changing color in 3% KOH.

Cells at the surface of the perithecial wall forming *textura epidermoidea.* Perithecial wall 15–20 μm wide, comprising a single region of nearly rectangular cells 5–7 × 2.5–3.5 μm, with walls ca. 1 μm thick. Perithecial papilla formed of chains of ± elliptic cells 5–7 × 2–4 μm, walls ca. 1 μm thick, cells at the surface of the papilla nearly circular in outline, 4–5 μm diam.; cells of the papilla becoming increasingly hyphal toward the ostiolar canal and merging with the periphyses.

Asci clavate, 40–50 × 5.5–8(–9) μm, apex simple; 8-spored, ascospores 2-seriate, completely filling each ascus. Ascospores ellipsoidal, (7.2–)7.9–9.2(–10) × ca. 3 μm, equally 2-celled, not constricted at the septum, smooth, colorless.

Habitat. On immersed ascomycetes, possibly also on *Nectria flavolanata* Berk. & Br.

Known Distribution. Brazil, known only from the type locality.

Etymology. This extremely small perithecial species is named in reference to Johnathan Swift's imaginary land of tiny characters.

Holotype. Brazil. Amazonas: Km 211 of Perimetral Norte hwy., Pico Rondon, vine forest, ca. 3 hr walk from FUNAI post, 01°31′N, 62°48′W, on palm midrib, possibly on *Linocarpon* sp., *Samuels 1138,* 24–26 May 1984 (INPA; isotype NY).

Additional Specimen Examined. BRAZIL. AMAZONAS: 0–3 km N of km 211 of Perimetral Norte hwy., Pico Rondon, lower vine forest; 01°32′N, 62°48′W, *Samuels 174A,* 4 Feb 1984 (INPA, NY).

Notes. *Nectriopsis lilliputia* is superficially similar to *N. septofusidiae,* also a parasite of immersed ascomycetes, but differs in having smaller ascospores and in the absence of dense mycelium. *Nectriopsis lilliputia* differs from *N. squamulosa* in having a simple ascal apex and in having somewhat shorter and narrower ascospores.

58. **Nectriopsis cordiae** (Rehm) Samuels, comb. nov. (Fig. 26).

≡*Sphaerostilbe cordiae* Rehm, Ann. Mycol. **6:** 486. 25 Nov 1908.

Anamorph. ?*Stilbella* sp.

Mycelium white, cottony, restricted to black colonies of a foliicolous ascomycete. Synnematous conidiomata scattered on leaf, up to 250 μm long with a single, terminal head of conidia, pale yellow. Conidiophores branching apically in an irregular fashion. Phialides 15–20 μm

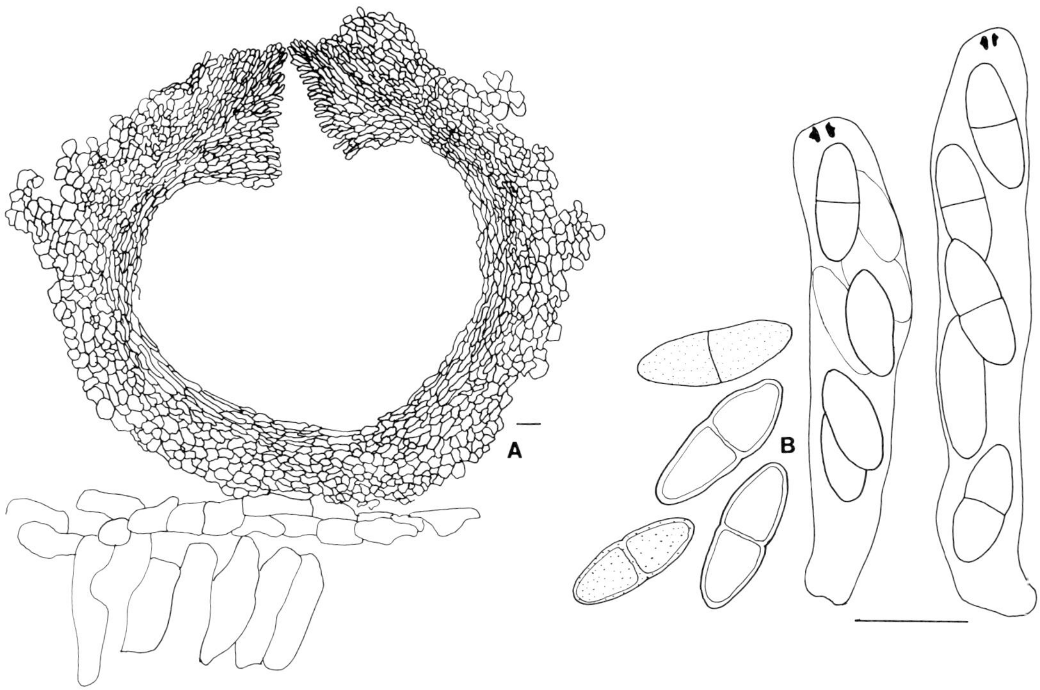

FIG. 26. *Nectriopsis cordiae.* **A.** Section through a mature perithecium. **B.** Asci and ascospores. Both drawn from the holotype. Lines = 10 μm.

long, tapering slightly from 2–3 μm wide at base to 2 μm at tip, smooth, straight; tip not thickened, unflared. Conidia fusiform with a subacute apex, (8–)8.5–12(–15) × 3–4 μm, base with a protuberant, flattened abscission scar, unicellular, smooth, colorless. Perithecia globse to obpyriform, 250–280 μm diam., nonpapillate, with white scales, collapsing by lateral pinching when dry; pale orange, not changing color in 3% KOH.

Cells at surface of perithecial wall circular to angular in outline, 7–12 μm across, walls 0.5–1 μm thick. Perithecial wall 25–30 μm wide, cells angular, 5–7 μm in greatest dimension, thin-walled; cells lining the locule flattened and compressed. Ostiolar region comprising ± vertically oriented, thin-walled cells ca. 5 μm long × 2 μm wide; ostiolar canal periphysate.

Asci broadly cylindrical, 50–80 × 10–12 μm, apex broad, with a ring; 8-spored, ascospores biseriate, completely filling each ascus. Ascospores elliptic-fusiform, (11–)12–15(–17) × (4–)4.7–5.5(–6) μm, equally 2-celled, slightly constricted at the septum, finely spinulose, colorless.

Habitat. On black fungus on leaves of *Cordia umbraculifera* DC.

Known Distribution. Brazil, known only from the type collection.

Holotype. Brazil. Pará: On leaves of *Cordia umbraculifera, C. F. Baker,* Jan 1908 (*Rehm: Ascomyceten 1782*: S!; isotype K!).

Notes. Synnemata were present only on the isotype found at S. The species has never been grown in pure culture. Although a link between this teleomorph and anamorph would not be inconsistent, it remains to be proven.

59. Nectriopsis epichloe (Spegazzini) Samuels, comb. nov. (Figs. 27, 28A, B).

≡*Nectria epichloe* Spegazzini, Anal. Soc. Ci. Argent. **19:** 39. Jan 1885.
=*Nectria epichloe* Spegazzini var. *rosea* Bresadola, Hedwigia **35:** 299. 1896.

Anamorph. *Myrothecium*-like.

Mycelium white, spreading somewhat beyond perithecia. Conidiomata poorly developed sporodochia, ≤500 μm diam., comprising ± loosely joined conidiophores. Conidiophores attaining a uniform height, 100 μm long, ± dichotomously

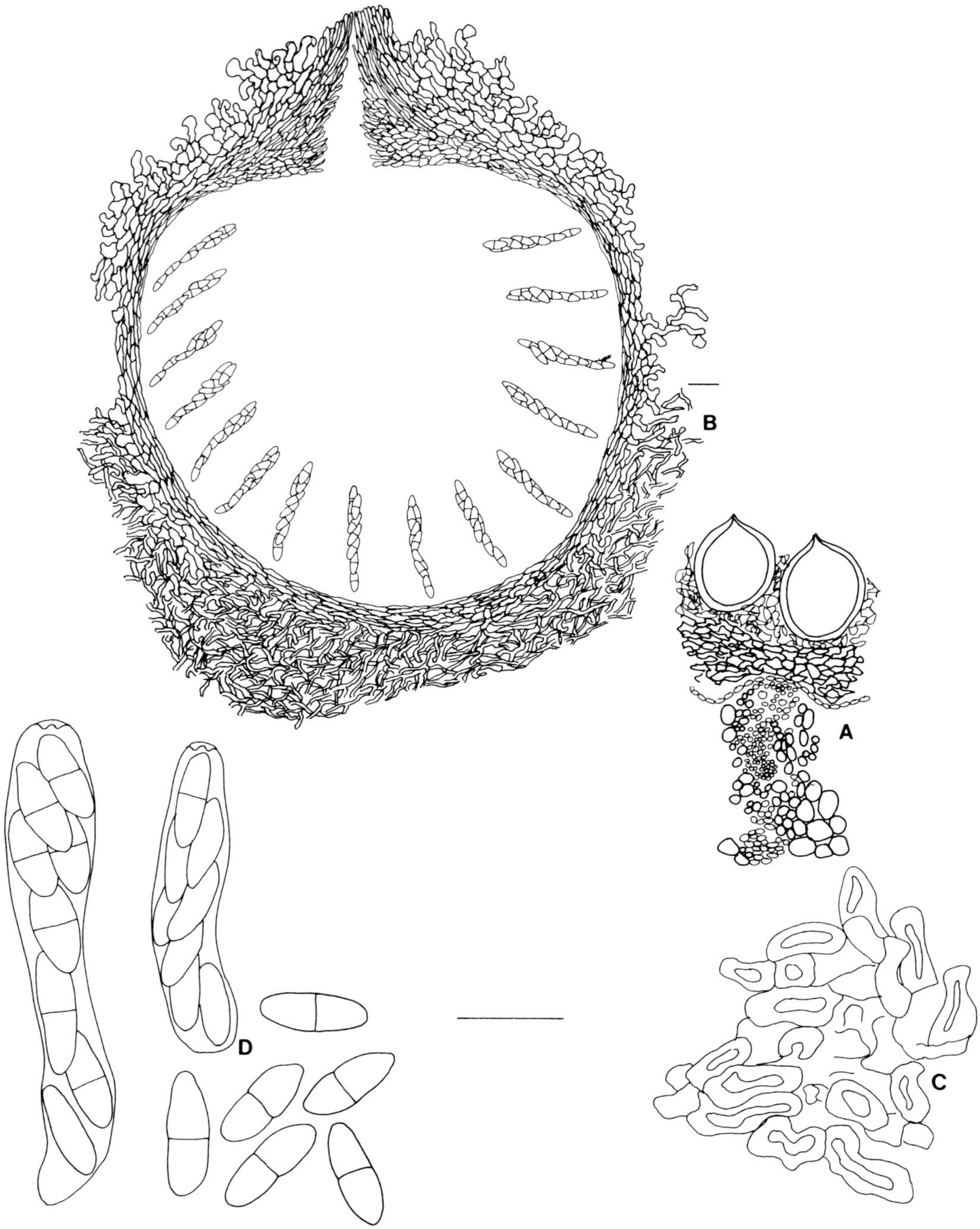

FIG. 27. *Nectriopsis epichloe*. **A.** Habit sketch of two perithecia on a grass culm (holotype). **B.** Section through a mature perithecium (holotype). **C.** Hair-like outgrowths from the perithecial wall (holotype of *Nectria epichloe* var. *rosea*). **D.** Asci and ascospores (holotype of *Nectria epichloe* var. *rosea*). Lines = 10 μm.

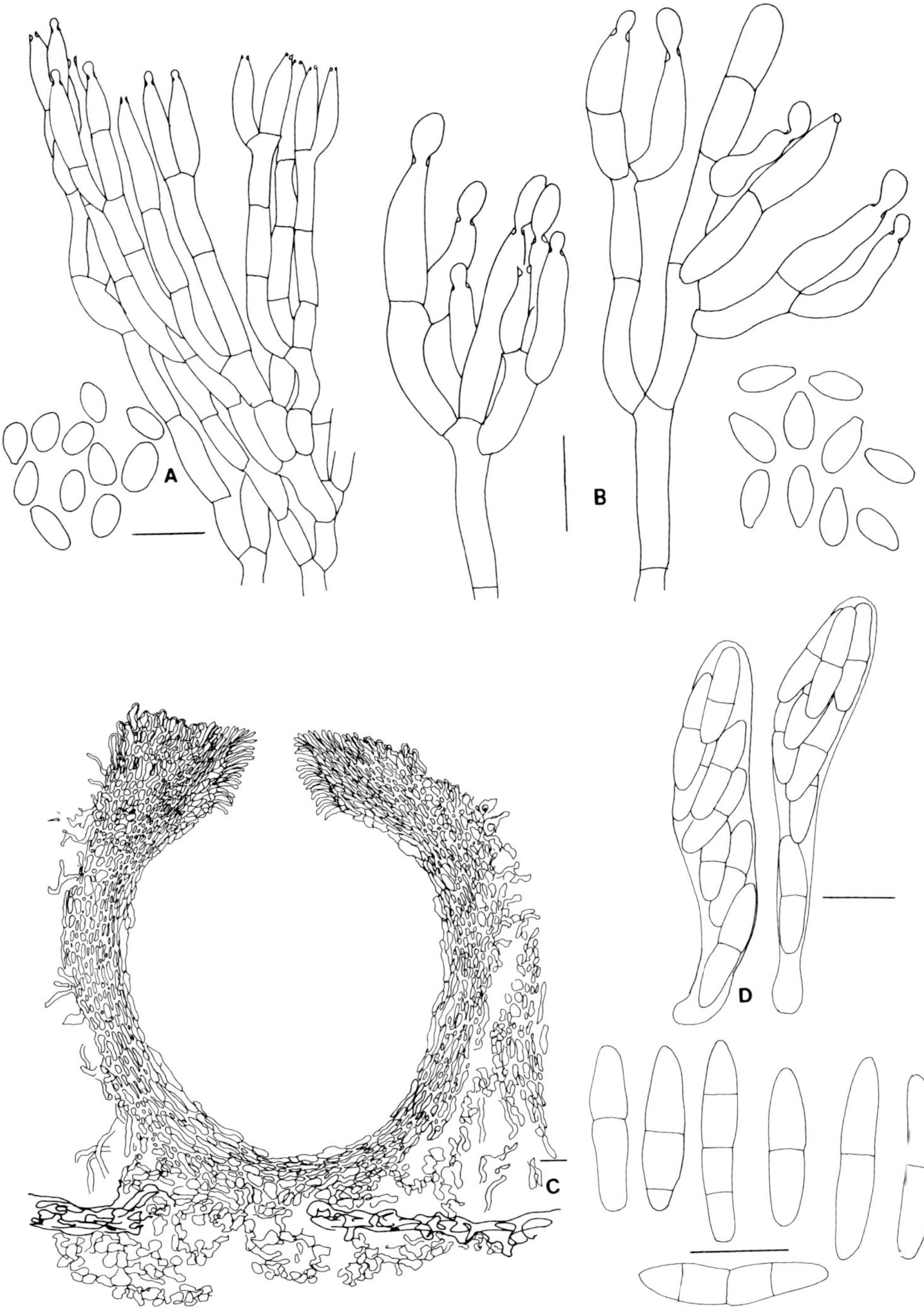

FIG. 28. A, B. *Nectriopsis epichloe.* **A.** Conidiophores and conidia from Brazil from nature (holotype of *Nectria epichloe* var. *rosea*). **B.** Conidiophores and conidia from U.S.A. (PDA, *Rogerson 79-207*). C, D. *Nectriopsis macroepichloe.* **C.** Section through a mature perithecium. **D.** Asci and ascospores. Both drawn from the holotype. Lines = 10 μm.

branched, each branch terminating in 1–2 phialides, smooth, pale green. Phialides cylindrical, (8–)10–17 × 2–3 μm, tip papillate to rounded, collarette slightly thickened, not flared, straight, smooth, pale green. Conidia ellipsoidal, (5–)6–8 × (2.5–)3–4(–5) μm, with or without a protuberant, flattened, ca. 0.5 μm wide basal abscission scar, unicellular, smooth, pale green, produced in basipetal succession and held in green slime masses.

Perithecia solitary to densely gregarious, superficial, nonstromatic, seated in or nearly covered by white mycelium, the aggregate of mycelium and perithecia 1–2 cm long, completely girdling the grass culm; perithecia globose, (200–)220–250(–280) μm diam., nonpapillate, smooth or with scattered warts, becoming cupulate or laterally pinched when dry; yellow to orange-brown, not changing color in 3% KOH.

Cells at the surface of the perithecial wall *textura epidermoidea,* <10 μm across, walls ca. 0.5 μm thick. Perithecial wall 15–20 μm wide, comprising a single region of elliptical to flattened cells 8–10 × 2–3 μm, walls ca. 0.5 μm thick; cells of the upper half of the perithecial wall producing short, unbranched, hyphal outgrowths, ≤20 μm long, walls ca. 1 μm thick, and many free ends; ends hooked or variously distorted or nearly globose, forming irregular, wart-like aggregates 60–100 μm across. Ostiolar region formed of narrow hyphal elements, hyphal elements merging with the periphyses.

Asci clavate, (34–)35–48(–51) × (5–)7–9(–10) μm, apex with a ring; 8-spored, ascospores biseriate, forming throughout the length of each ascus. Ascospores fusiform, (8.5–)9.3–11.5(–13) × (3–)3.3–4(–4.5) μm, equally 2-celled, slightly constricted at the septum, smooth to slightly spinulose, colorless.

Characteristics in Culture. Colonies grown 5 weeks at ca. 20°C, diffused daylight. PDA: ca. 2 cm diam., flat, felty, opaque, pink. CMD: 3 cm diam., flat, felty, translucent, pale orange. Conidiophores forming on both media, held in sporodochial aggregates, up to 100 μm long, variable in length, irregularly branched, each branch terminating in 1–2 phialides. Phialides cylindrical, (7–)14–19(–29) × 3–4 μm, straight, smooth, collarette thickened, not flared. Conidia ellipsoidal to clavate, (6–)6.7–8.4(–9.5) × (2–)3–3.5 μm, with a protuberant, flat, 1 μm wide basal abscission scar, unicellular, smooth, pale green, held in green masses of slime.

Habitat. On stromata of *Epichloe* sp. and *Balansia* sp. on grass culms.

Known Distribution. Brazil, U.S.A. (Georgia).

Holotype. Brazil. Guarapí, on *Andropogon* sp., *Balansa 3432,* Dec 1882 (*Balansa, Plantes du Paraguay 3432*: LPS!; isotype NY!).

Additional Specimens Examined. USA. GEORGIA: Newton County, on stroma of *Balansia henningsiana* (Möll.) Diehl, *Bacon,* Oct 1978 (*Rogerson 79-207*: NY, PDD 40125). BRAZIL. SANTA CATARINA: Blumenau, on *Andropogon, Möller 28a* (S, herb. Bresadola, holotype of *Nectria epichloe* var. *rosea*).

Notes. Although described as a parasite of grasses, *Nectriopsis epichloe* is clearly a parasite of *Balansia* and *Epichloe* species that are themselves parasites of the grasses. In their study of the anamorph of *Epichloe typhina* (Fries) Tul., Morgan-Jones & Gams (1982) reported finding green conidia in a specimen of *Sphacelia typhina aeruginosa* Sacc. (*Therry, Cryptog. de Lyonnais 1881*). They concluded that *S. typhina aeruginosa* was *Epichloe typhina* parasitized by *Nectria pityrodes.* It is more likely that the conidia that they observed were of *Nectriopsis epichloe*; *Nectria pityrodes* is a lignicolous species but does have a *Myrothecium* anamorph.

Nectriopsis epichloe closely resembles *N. macroepichloe,* from which it differs in having smaller ascospores.

60. Nectriopsis macroepichloe Samuels, sp. nov. (Fig. 28C, D).

Nectriopsidi epichloi (Speg.) Samuels similis, sed ascosporis 15.3–19.2(–23) × 3–4 μm. In *Balansia* crescens.

Status anamorphicus ignota.
Holotypus. Venezuela (BPI: *Chardon 2785*).
Anamorph. None known.

Mycelium thin, white, spreading; hyphae 2–3 μm wide, wall ca. 0.5 μm thick, smooth. Perithecia densely gregarious, forming in a continuous layer over surface of host stroma, superficial, nonstromatic, globose, 190–250 μm diam., nonpapillate, smooth; pale yellow to nearly white, not changing color in 3% KOH.

Cells at surface of perithecial wall forming *textura epidermoidea,* ≤10 μm across, wall <0.5 μm thick. Perithecial wall 20–25 μm wide, comprising two regions; outer region 10–15 μm wide, cells elliptic in section, 5–7 × 2–3 μm, walls <0.5 μm thick; inner region 10–15 μm wide, cells flattened, ca. 10 × 2 μm, walls <0.5 μm thick. Perithecial apex not anatomically distinct from the wall below but cells around the ostiolar opening elliptic, and some cells at wall surface growing out as hairs to a distance of ca. 10 μm; ostiolar canal periphysate.

Asci clavate, (42–)45–62(–75) × (7–)8–12 μm, apex simple; 8-spored, ascospores 2- to multiseriate, completely filling each ascus. Ascospores oblong to ellipsoidal, 15.3–19.2(–23) × 3–4 μm, equally 2-celled with two additional septa often forming after discharge, smooth, colorless.

Habitat. On stroma of *Balansia.*

Known Distribution. Brazil, Venezuela.

Etymology of the Specific Epithet. Refers to the general similarity to *Nectriopsis epichloe* but differing in large ascospores.

Holotype. Venezuela. Carabobo: Mts. above Hacienda Cura, on stroma of *Hypocrella cyperacearum* (Berk. & Curt.) Sacc. (=*Balansia cyperacearum* (Berk. & Curt.) Diehl.) on *Scleria bracteata* Cav., *Chardon 2785,* 8 Jan 1939 (BPI).

Additional Specimen Examined. BRAZIL. on *Balansia* sp., *A. S. Muller 2038,* 1940 (BPI, microscope slide only).

Notes. There is reason to doubt whether this species has been found in Brazil. The specimen *Muller 2038* is *Dothichloe nigricans* (Speg.) Seaver *in* Stevenson (=*Balansia strangulans* (Mont.) Diehl f. *discoidea* (Henn.) Diehl) on *Chloris polydactyla* (L.) Swartz collected in Venezuela on 13 Nov 1937. The Brazilian collection cited above was found on the same herbarium sheet with *Chardon 2785*; both were identified by Diehl as *Nectria fuckelii* Sacc. The *Muller* specimen was in a packet with a label indicating that the specimen was found in Brazil. The packet bore the annotation, presumably by Diehl, "Specimen not found" and contains three microscope slides, all of *Nectriopsis macroepichloe.*

Acknowledgments

This research has been supported by grants to the author from the American Philosophical Society, the National Geographic Society (NGS 2736-83, 3068-85) and the National Science Foundation (BSR 8500236). Additional support was provided by Projeto Flora Amazônica (NSF grants INT 77-17704, DEB 8106632 to the New York Botanical Garden) and Project Wallace (the Royal Entomological Society and the Indonesian Institute of Sciences). I am grateful to the directors of the following herbaria for the loan of specimens: BPI, FH, HMAS, IACM, IMI, K, LPS, PACA, PDD, S, TNS, W. Drs. O. Eriksson and Jing-zhu Yue called *Chromocreopsis epinectria* to my attention. Dr. Rupert Barneby corrected the Latin diagnoses. Dr. Richard Harris identified the lichen substrates. Dr. Clark Rogerson kindly read the various drafts of this article and was unstintingly sharing of his taxonomic insights. To these people also I am very grateful.

Literature Cited

Barr, M. E. 1980. On the family Tubeufiaceae (Pleosporales). Mycotaxon **12:** 137–167.

Batista, A. C., G. E. P. Peres & D. J. M. Poroca. 1967. *Pseudonectria maranhensis* n. sp. e outros Nectriaceae hiperparasiticos. Atas Inst. Micol. Recife **5:** 131–137.

Booth, C. 1959. Studies of pyrenomycetes: IV. *Nectria* (Part I). Mycol. Pap. **73:** 1–115.

———. 1966. The genus *Cylindrocarpon.* Mycol. Pap. **104:** 1–56.

Cauchon, R. & G. B. Ouellette. 1964. Association of *Stromatocrea cerebriforme* and *Hypocreopsis* species with *Hymenochaete* species. Mycologia **56:** 453–455.

Crosby, T. K., J. S. Dugdale & J. C. Watt. 1976. Recording specimen localities in New Zealand: An arbitrary system of areas and codes defined. New Zealand J. Zool. **3:** 69.

Crouan, P. L. & H. M. Crouan. 1867. Florule du Finistère. Paris.

Dennis, R. W. G. 1978. British Ascomycetes. Cramer, Vaduz.

Doi, Y. 1967. A revision of Hypocreales with cultural observation II. On *Mycocitrus phyllostachydis* (Syd.) Doi, a perfect state of *Cephalosporium.* Bull. Nat. Sci. Mus., Tokyo **10:** 31–36.

———. 1977. Revision of the Hypocreales with cultural observations X. Two new species of the genus *Hypocreopsis.* Bull. Nat. Sci. Mus., Ser. B (Bot.) **3:** 99–104.

———. 1978. The Hypocreales (Fungi, Ascomycetes) of the Bonin Islands (2). Mem. Nat. Sci. Mus. Tokyo **11:** 17–20.

Domsch, K. H., W. Gams & T.-H. Anderson. 1980. Compendium of soil fungi. Vol. 1. Academic Press, New York.

Gams, W. 1971. *Cephalosporium*-artige Schimmelpilze (Hyphomycetes). G. Fischer, Stuttgart.

——— **& A. van Zaayen.** 1982. Contribution to the taxonomy and pathogenicity of fungicolous *Verticillium* species. I. Taxonomy. Netherlands J. Pl. Pathol. **88:** 57–78.

Gerlach, W. & H. Nirenberg. 1982. The genus *Fusarium*—A pictorial atlas. Mitt. Biol. Bundesanst. Land-Forstw. Berlin-Dahlem **209:** 1–406.

Hansford, C. G. 1946. The foliicolous ascomycetes, their parasites and associated fungi especially as illustrated by Uganda specimens. Mycol. Pap. **15:** 1–240.

Kirk, P. M. 1983. New or interesting microfungi X. Hyphomycetes on *Laurus nobilis* leaf litter. Mycotaxon **18:** 259–298.

Kobayashi, Y. & Y. Doi. 1969. On *Phaeocreopsis hypoxyloides* (Speg.) Sacc. & Syd. Mem. Nat. Sci. Mus., Tokyo **2:** 51–56.

Maire, R. 1911. Remarques sur quelques Hypocréacées. Ann. Mycol. **9:** 315–325.

Morgan Jones, G. & W. Gams. 1982. Notes on Hyphomycetes. XLI. An endophyte of *Festuca arundinacea* and the anamorph of *Epichloe typhina,* new taxa in one of two new sections of *Acremonium.* Mycotaxon **15:** 311–318.

Müller, E. & J. A. von Arx. 1962. Die Gattungen der didymosporen Pyrenomyceten. Beitr. Kryptogamenfl. Schweiz **11(2):** 1–922.

Petrak, F. 1952. Ergebnisse einer Revision der Grundtypen verschiedener Gattungen der Askomyceten und Fungi Imperfecti. III. Sydowia **6:** 336–343.

Pfister, D. H. 1976. *Calloriopsis* and *Micropyxis*: Two discomycete genera in the *Calloriopsidae* trib. nov. Mycotaxon **4:** 340–346.

Pirozynski, K. A. 1976. Notes on hyperparasitic Sphaeriales, Hypocreales and 'Hypocreoid Dothideales.' Kew Bull. **31:** 595–610.

Rifai, M. A. 1969. *Sarawakus* Lloyd, a genus of the pyrenomycete family Hypocreaceae. Reinwardtia **7:** 561–578.

Rogerson, C. T. 1970. The hypocrealean fungi. Mycologia **62:** 865–910.

——— **& G. J. Samuels.** 1985. Species of *Hypomyces* and *Nectria* occurring on discomycetes. Mycologia **77:** 763–783.

Rossman, A. Y. 1979. A preliminary account of the taxa described in *Calonectria.* Mycotaxon **8:** 485–558.

———. 1983. The phragmosporous species of *Nectria* and related genera. Mycol. Pap. **150**: 1–164.

———. 1987. The Tubeufiaceae and similar Loculoascomycetes. Mycol. Pap. **157:** 1–71.

Samson, R. A. 1974. *Paecilomyces* and some allied Hyphomycetes. Stud. Mycol. **6:** 1–119.

Samuels, G. J. 1971. The taxonomic position of the genus *Nectriopsis* Maire (Fungi, Ascomycetes, Hypocreales). Unpublished Ph.D. Dissertation. Columbia University, New York.

———. 1973. The myxomyceticolous species of *Nectria.* Mycologia **65:** 401–420.

———. 1976a. A revision of the fungi formerly classified as *Nectria* subgenus *Hyphonectria.* Mem. N.Y. Bot. Gard. **26(3):** 1–126.

———. 1976b. Perfect states of *Acremonium.* The genera *Nectria, Actiniopsis, Ijuhya, Neohenningsia, Ophiodictyon,* and *Peristomialis.* New Zealand Jour. Bot. **14:** 231–260.

———. 1978. Some species of *Nectria* having *Cylindrocarpon* imperfect states. New Zealand Jour. Bot. **16:** 73–82.

——— **& A. Y. Rossman.** 1979. Conidia and classification of the nectrioid fungi. *In* B. Kendrick, ed., The whole fungus. Vol. 1. National Museums of Canada, Ottawa.

——— **& K. A. Seifert.** 1987. Taxonomic implications of variation among hypocrealean anamorphs. *In* J. Sugiyama, ed., Pleomorphic fungi: The diversity and its taxonomic implications. Kodansha, Tokyo.

Seifert, K. A. 1985. A monograph of *Stilbella* and some allied Hyphomycetes. Stud. Mycol. **27:** 1–235.

Spegazzini, C. 1919. Fungi costaricensis nonnulli. Bol. Acad. Nac. Cienc. Córdoba **23:** 541–593.

Starbäck, K. 1899. Ascomyceten der ersten regnellischen Expedition. Kongl. Svenska Vetenskapsakad. Handl. Bih. **25(iii: 1):** 1–68.

Strid, A. 1967. *Hypocreopsis lichenoides* (Tode ex Fr.) Seaver, a rare Ascomycete. Svensk Bot. Tidsk. **61:** 70–87.

Teng, S. C. 1935. Supplementary notes on Ascomycetes from China. Sinensia **6:** 185–218.

Tulloch, M. 1972. The genus *Myrothecium* Tode ex Fr. Mycol. Pap. **130:** 1–42.

Index

Fungi and their substrates are included. New names and combinations are in **boldface**, and synonyms are in *italics*. Page numbers in **boldface** indicate primary page references. Page numbers with an asterisk (*) indicate pages with illustrations.

GARY J. SAMUELS

Dr. Gary J. Samuels was born in Grove City, Pennsylvania in 1944. He obtained his B.Sc. Degree in Botany from the Pennsylvania State University in 1966, and then went to the New York Botanical Garden where he was a graduate fellow, matriculating in the Ph.D. program in Biology at Columbia University. He earned his M.A. Degree in 1968 and his Ph.D. Degree in 1971 as a student of Dr. Clark T. Rogerson. Dr. Samuels was a scientist with the New Zealand Department of Scientific and Industrial Research, Plant Diseases Division in Auckland from 1973 until 1985. During that time he studied the fungi that cause plant disease, as well as the ascomycetes occurring naturally in New Zealand. In 1977–1978 he was a Visiting Guest Researcher in the laboratory of Prof. Emil Müller, Eidg. Technische Hochschule, Institut für Spezielle Botanik, Zürich where he studied the ascomycetes of Brazil. In 1985 he joined the staff of the New York Botanical Garden as a Research Associate, where he has been supported by grants from the National Science Foundation to study the Hypocreales of tropical America. Dr. Samuels has carried out field work in Brazil, French Guiana, Guyana, Indonesia, Jamaica, Japan, New Zealand, and Venezuela. His publications include the multivolume Index to the Mycological Writings of Franz Petrak in addition to approximately eighty scientific articles that describe the ascomycetes of Indonesia, New Zealand, and the neotropics.

MEMOIRS OF THE NEW YORK BOTANICAL GARDEN

The Botany of the Guayana Highland—Parts I-XIII
by Bassett Maguire and Collaborators

I.	Mem. N. Y. Bot. Gard. 8(2): 87–160. 1953. (Combretaceae, Compositae, Eriocaulaceae, Leguminosae, Malpighiaceae, Melastomataceae, Polygalaceae, Rutaceae)	$10.00
II.	Mem. N. Y. Bot. Gard. 9(3): 235–392. 1957. (Annonaceae, Begoniaceae, Bignoniaceae, Bromeliaceae, Compositae, Droseraceae, Eriocaulaceae, Gramineae, Haemadoraceae, Leguminosae, Musci, Piperaceae, Polygalaceae)	$15.00
III.	Mem. N. Y. Bot. Gard. 10(1): 1–156. 1958. (Apocynaceae, Bignoniaceae, Combretaceae, Guttiferae, Leguminosae, Melastomataceae, Myrtaceae, Rapateaceae, Xyridaceae)	$15.00
IV.	Mem. N. Y. Bot. Gard. 10(2): 1–37. 1960. (Musci)	$10.00
IV.	Mem. N. Y. Bot. Gard. 10(4): 1–87. 1960. (Acanthaceae, Annonaceae, Aquifoliaceae, Bromeliaceae, Guttiferae, Linaceae, Marantaceae, Melastomataceae, Ochnaceae, Rutaceae, Xyridaceae)	$10.00
V.	Mem. N. Y. Bot. Gard. 10(5): 1–278. 1960. (Bambusoideae, Bromeliaceae, Guttiferae, Lauraceae, Melastomataceae, Piperaceae, Rubiaceae, Xyridaceae)	$17.50
VI.	Mem. N. Y. Bot. Gard. 12(3): 1–285. 1965. (Aquifoliaceae, Cyperaceae, Euphorbiaceae, Gramineae, Lauraceae, Rapateaceae, Rubiaceae)	$17.50
VII.	Mem. N. Y. Bot. Gard. 17(1): 1–439. 1967. (Abolbodaceae, Bombacaceae, Compositae, Cyperaceae, Euphorbiaceae, Labiatae, Lentibulariaceae, Polypodiaceae, Rubiaceae)	$35.00
VIII.	Mem. N. Y. Bot. Gard. 18(2): 1–290. 1969. (Annonaceae, Apocynaceae, Bromeliaceae, Cyperaceae, Gramineae, Malpighiaceae, Melastomataceae, Myrtaceae, Quiinaceae, Schizaeaceae, Velloziaceae)	$20.00
IX.	Mem. N. Y. Bot. Gard. 23: 1–832. 1972. (Bonnetiaceae, Caryocaraceae, Clusiaceae, Cycadeaceae, Cyclanthaceae, Gramineae, Meliosmaceae, Melastomataceae, Palmae, Piperaceae, Polygalaceae, Polypodiaceae, Rubiaceae, Sapotaceae, Tetrameristaceae)	$50.00
X.	Mem. N. Y. Bot. Gard. 29: 1–288. 1978. (Bonnetiaceae, Bignoniaceae, Cyatheaceae, Ericaceae, Icacinaceae, Saccifoliaceae, Sarraceniaceae, Styracaceae, Symplocaceae, Tiliaceae)	$25.00
XI.	Mem. N. Y. Bot. Gard. 32: 1–391. 1981. (Tepuianthaceae, Malpighiaceae, Dipterocarpaceae, Convolvulaceae, Ebenaceae, Lissocarpaceae, Gentianaceae)	$40.00
XII.	Mem. N. Y. Bot. Gard. 38: 1–84. 1984. (Hymenophyllopsidaceae, Hymenophyllaceae, Araliaceae)	$16.00
XIII.	Mem. N.Y. Bot. Gard. 50: 1–00. In press. (Gentianaceae, Ochnaceae, Melastomataceae, Dioscoreaceae, Nyctaginaceae, Monimiaceae, Capparaceae, Cunoniaceae, Rhamnaceae, Rutaceae)	$

Postage and Handling:

U.S. Orders:	$2.00 + 3% of subtotal.
Non-U.S. Orders:	$3.00 + 4% of subtotal. Payment in U.S. currency drawn on a U.S. bank. Thank you.

Order from: MEMOIRS, Scientific Publications Department, The New York Botanical Garden, Bronx, NY 10458 USA.

Memoirs of The New York Botanical Garden*

VOLUME		PRICE
29	**Maguire, B. and Collaborators.** 1978. The botany of the Guayana Highland—Part X. (288 pp.)	$25.00
30	**Irwin, H. S. & R. C. Barneby.** 1978. Monographic studies in *Cassia* (Leguminosae—Caesalpinioideae). III. Section *Absus* and *Grimaldia.* (300 pp.)	$12.50
31(1)	**Grear, J. W.** 1978. A revision of the New World species of *Rhynchosia* (Leguminosae—Faboideae). (168 pp.)	$15.00
31(2)	**Becker, K. M.** 1979. A monograph of the genus *Lasianthea* (Asteraceae). (64 pp.)	$ 6.50
32	**Maguire, B. and Collaborators.** 1981. The botany of the Guayana Highland—Part XI. (391 pp.)	$40.00
33	**Carpenter, S. E.** 1981. Monograph of *Crocicreas* (Ascomycetes, Helotiales, Leotiaceae). (290 pp.)	$35.00
34	**Furtado, J. S.** 1981. Taxonomy of *Amauroderma* (Basidiomycetes, Polyporaceae). (109 pp.)	$17.50
35	**Irwin, H. S. & R. C. Barneby.** 1982. The American Cassiinae. A synoptical revision of Leguminosae tribe Cassieae subtribe Cassiinae in the New World. (918 pp.)	$100.00
36	**Funk, V. A.** 1982. The systematics of *Montanoa* (Asteraceae, Heliantheae). (133 pp.)	$21.00
37	**Thomas, W. W.** 1984. The systematics of *Rhynchospora* section *Dichromena.* (116 pp.)	$21.00
38	**Maguire, B. and Collaborators.** 1984. The botany of the Guayana Highland—Part XII. (84 pp.)	$16.00
39	**Ertter, B.** 1986. The *Juncus triformis* complex. (90 pp.)	$17.50
40	**Halling, R.** 1986. An annotated index to species and infraspecific taxa of Agaricales and Boletales described by William A. Murrill. (140 pp.)	$19.25
41	**Leuenberger, B. E.** 1986. *Pereskia* (Cactaceae). (141 pp.)	$25.00
42	**Mitchell, J. D. & S. A. Mori.** 1987. The cashew and its relatives (*Anacardium*: Anacardiaceae). (76 pp.)	$17.50
43	**Pipoly, J. J. III.** 1987. A systematic revision of the genus *Cybianthus* subgenus *Grammadenia* (Myrsinaceae). (76 pp.)	$17.50
44	**Mori, S. A. and Collaborators.** 1987. The Lecythidaceae of a lowland neotropical forest: La Fumée Mountain, French Guiana (190 pp.)	$37.00
45	**Buck, W. R., Editor.** 1987. Bryostephane Steereana: A collection of bryological papers presented to William Campbell Steere on the occasion of his 80th birthday. (778 pp.)	$130.00
46	**Mickel, L. T. and J. M. Beitel.** 1988. Pteridophyte flora of Oaxaca, Mexico. (568 pp.)	$
47	**Park, C.-W.** 1988. Taxonomy of *Polygonum* section *Echinocaulon* (Polygonaceae). (82 pp.)	$
48	**Samuels, G. J.** In press. Fungicolous, lichenicolous, and myxomyceticolous species of *Hypocreopsis, Nectriopsis, Nectria, Peristomialis,* and *Trichonectria.* (000 pp.)	$
49	**Renner, S. S.** In press. Systematic studies in the Melastomataceae: *Bellucia, Loreya,* and *Macairea.* (000 pp.)	$
50	**Maguire, B. and Collaborators.** In press. The botany of the Guayana Highland—Part XIII. (000 pp.)	$

Postage and Handling:

U.S. ORDERS: $2.00 + 3% of subtotal.

NON-U.S. ORDERS: $3.00 + 4% of subtotal. Payment in U.S. currency drawn on a U.S. bank. Thank you.

Order from: MEMOIRS, Scientific Publications Department, The New York Botanical Garden, Bronx, NY 10458 U.S.A.

* Volumes 1–28 have been published. Write to the address above for information as to their status and prices.